Table of Contents

Page

LIST OF ILLUSTRATIONS iii

FIRE FROM THE SKY 1

THE THREAT TO PLANET EARTH 7

    Size vs Hazard... 9

        Category 1: 10-100m Diameter Impactors 10

        Category 2: 100m-1km Diameter Impactors 11

        Category 3: 1km-5km diameter impactors 12

    Unique Aspects of Comets... 14

    Risk 14

        Probability of Occurrence 15

        Risk Acceptance... 18

        A Signal 23

        Unnatural Power 25

        What Does It Mean? 26

POTENTIAL SOLUTIONS.. 31

    Surveillance and Control.. 32

    Mitigation.. 34

    Cost 38

    Legality of Planetary Defense.. 39

        Outer Space Treaty... 40

        Nuclear Test Ban Treaty 41

        Anti-Ballistic Missile Treaty.. 43

        The Legal Road Ahead 43

CURRENT PROGRAMS 46

    Surveillance Programs 46

    Characterization Programs.. 47

        Ongoing Programs 48

Future Programs...48
Mitigation Programs 50
The Air Force 51

TODAY'S ISSUES 54
What We Know...54
At Issue 55
Is the asteroid/comet threat of sufficient concern to merit taking action?............55
Why hasn't action been taken already? 58
Who is responsible for taking action?..62
Where will the money come from?...64
What must be done now? 66

RECOMMENDATIONS 69
National Policy..70
Commitment 71
International Mobilization 72
Focus...73
Organization..75
Defense Mission..76
System Test 78

OUR LEGACY FOR THE FUTURE 82

GLOSSARY..84

BIBLIOGRAPHY 85

List of Illustrations

Figure 1. Aerial photo taken in the Tunguska region 30 years after the event. The fallen trees indicate the direction of the shock wave. 2

Figure 2. Meteor Crater in Arizona 2

Figure 3. Manicouagan Crater, Canada. Viewed from the Space Shuttle.......................... 3

Figure 4. Earth Orbit Vs 100 Largest Earth-Crossing Asteroids 16

Chapter 1

Fire From the Sky

I take my chances, I take my chances every chance I get.

— Mary-Chapin Carpenter

On the morning of June 30, 1908, a fireball cascaded down the Siberian sky and exploded with 2,000 times the force of the nuclear blast that devastated Hiroshima, Japan. Weighing some 100,000 metric tons, the cosmic missile cut into the atmosphere at about a 30° angle above the horizon and an azimuth of 110° out of the southeast. Observers described a fiery tail some 800 kilometers long. At an altitude of about 6 kilometers the object shattered in a rapid series of bursts and vaporized, felling trees in a radial pattern over an area 2,150 square kilometers and incinerating a central area twice that size.[1]

The shock wave from this event, centered over the Tunguska region of Siberia, was sufficiently powerful that it circumnavigated the globe, being recorded on successive days at Potsdam, Germany.[2] (Figure 1) Although it is unknown whether this object was a comet or an asteroid whose orbit intersected Earth's, it is fortunate that it exploded in the atmosphere without impacting the ground, and that it exploded over one of the most sparsely populated areas on the planet. Had this object encountered Earth three hours later, Moscow would have been leveled,[3] killing millions and changing Russian and world history forever. A unique event? Hardly. In fact, Earth history is replete with impacts of other-worldly objects. Many Americans are familiar with the meteor crater in Arizona, which is 4,000 feet across (Figure 2). It was created by an object

Figure 1. Aerial photo taken in the Tunguska region 30 years after the event. The fallen trees indicate the direction of the shock wave. Source: University of Bologna.

Figure 2. Meteor Crater in Arizona

approximately 60 meters in diameter, weighing several million tons which impacted

Earth traveling at approximately 10 miles per second.[4] Nearly all who visit the crater consider it a major event in geologic *history*, however, most regard it largely as a curiosity rather than coming to the realization that it is hard evidence of the reality of a *continuing* threat. But although this crater is impressive in its size and the impactor that created it powerful, they are actually extremely small in the historical record of Earth's encounters with other cosmic players. For example, an object impacted Quebec 214 million years ago creating the 100 kilometer wide Manicouagan Crater. (Figure 3) Seventy million years later, another object created a 22 kilometer crater in Australia; 65 million years ago, impact with an asteroid 10 kilometers in diameter left a crater 180 kilometers wide off the coast of the Yucatan peninsula.[5] One may view these examples as ancient history, but they're not the end of the story—they serve as portentous evidence

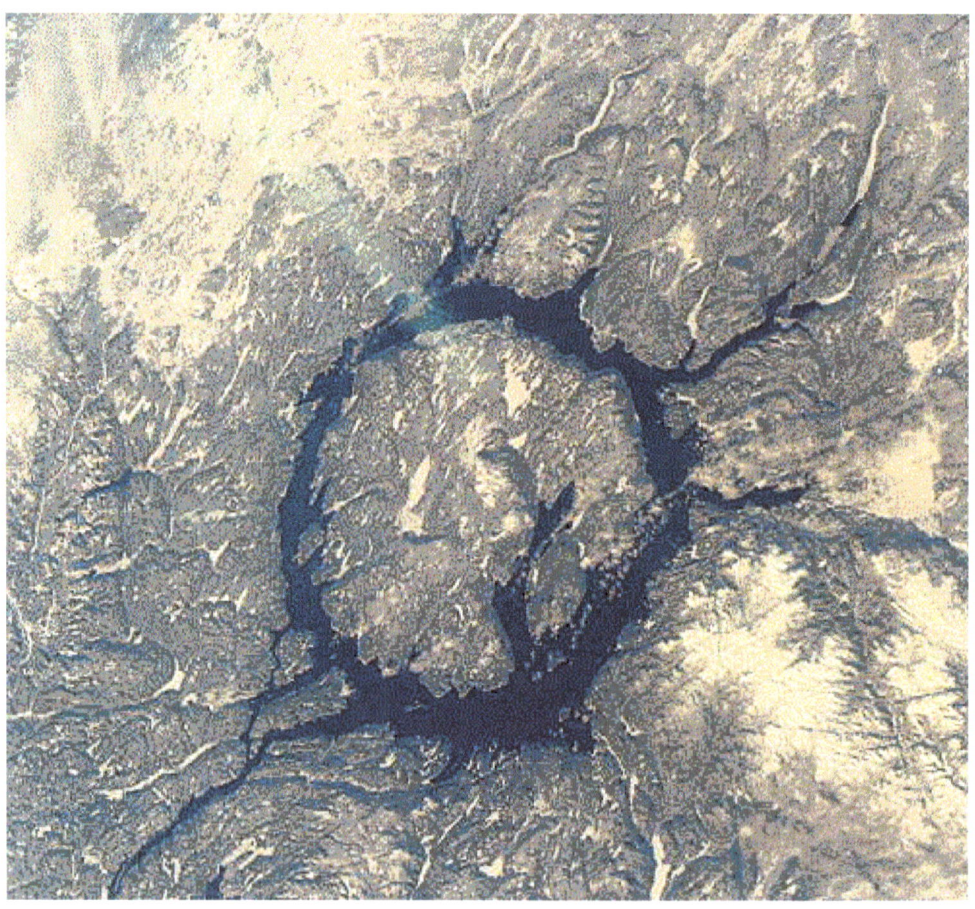

Figure 3. Manicouagan Crater, Canada. Viewed from the Space Shuttle. Source: NASA

of Earth's future.

Between 16 and 22 July 1994, 20 nuclei of Periodic Comet Shoemaker-Levy 9 impacted Jupiter with spectacular visible and electromagnetic effects. The size of the fragments is not known conclusively, but was estimated to range between 100 meters and 4 kilometers.[6] Had one of these fragments impacted Earth instead, the results would have been catastrophic, likely far worse than the 1908 "Tunguska" destruction.[7] Closer to home, on 22 March 1989 an asteroid designated 1989FC with a diameter of between 200 and 400 meters[8] passed closely by Earth with zero warning time—this asteroid had never been cataloged before.[9] On 8 December 1992, Asteroid Toutatis missed Earth by a scant 2 lunar distances—a negligible distance in celestial terms. Had this 4 kilometer object struck Earth, it would have done so with the force of 9 million megatons of TNT, more than all the nuclear weapons in existence combined.[10] On 1 February 1994, a small meteor traveling over 33,000 miles per hour exploded in the atmosphere off the coast of New Guinea. The blast energy was equivalent to 11 kilotons of TNT, and the explosion was described as being as bright as the sun.[11] These historical and recent events point to a significant, growing concern that our planet is subject to periodic attack by primarily comets and asteroids. Although the subject evokes a certain "giggle factor," this is due to lack of information, misinformation, or lack of personal knowledge of a recent catastrophe. Recorded human history is well versed in death and destruction associated with volcanoes, hurricanes, tornadoes, and similar natural disasters; we all have either experienced such disaster first hand or are painfully aware of the destruction wrought by them through personal contacts or credible news reports. In contrast, however, the last widespread loss of life due to a cosmic impact occurred in China in 1490 when approximately 10,000 people were killed—hardly in the experience base of our society today.[12] The Tunguska impact was certainly recent and disastrous, but not in human or economic terms. It is still unknown if anyone died as a result of the Tunguska impact.[13]

Perhaps we should consider ourselves lucky—lucky that the Tunguska destruction occurred in one of the most isolated and sparsely inhabited areas on Earth, and lucky that 1989FC and Toutatis were "near misses" rather than cataclysmic "hits." While I firmly believe in luck, I also know that all luck runs out. The central questions this paper will address are whether the threat posed by Earth-crossing asteroids and comets is significant enough to warrant taking action; if so, what should that action be? We must view the Tunguska explosion and the asteroid near misses as clarion calls to action. Then, rather than heaving a sigh of relief or reveling in thousands of years of blissful ignorance, we must recognize the existence of a very real threat to all mankind. I contend we must rapidly develop the means to defend ourselves, our civilization, our species, and our planet from the natural threat of impact from near-Earth celestial objects. To reach this conclusion, I will review the scope of the threat posed by near-Earth objects, the hazard and the risk; discuss the range of potential solutions; review the paucity of ongoing programs; illuminate the immediate planetary defense issues; and make recommendations for near term actions to be taken at the national level and by the Air Force.

Notes

[1] Roy A. Gallant, "Journey to Tunguska," *Sky & Telescope* 87, no. 6 (June 1994): 38.

[2] Larry D. Bell, William Bender, and Michael Carey, "Planetary Asteroid Defense Study: Assessing the Responding to the Natural Space Debris Threat," research paper ACSC/DR/225/95-04 (Maxwell AFB, Ala.: Air Command and Staff College, 1995), 58.

[3] John M. Urias et al., "Planetary Defense: Catastrophic Health Insurance for Planet Earth," research paper submitted to Air Force 2025, October 1996, n.p.; on-line, Internet, 23 September 1997, available from http://www.au.af.mil/au/2025volume3/chap16/

Notes

v3c16-1.htm

[4] Clark R. Chapman and David Morrison, *Cosmic Catastrophes* (New York: Plenum Press, 1989), 14, 18.

[5] Rosario Nici and Douglas Kaupa, "Planetary Defense: Department of Defense Cost for the Detection, Exploration, and Rendezvous Mission of Near-Earth Objects," *Airpower Journal* XI, no. 2 (Summer 1997): 94.

[6] John R. Spencer and Jacqueline Mitton, ed., *The Great Comet Crash: The Impact of Comet Shoemaker-Levy-9 on Jupiter* (Cambridge, UK: Cambridge University Press, 1995), vii, 97-99.

[7] Ibid., 106.

[8] "Closest Approaches to the Earth by Minor Planets," *International Astronomical Union Minor Planet Center.* N.d., n.p.; on-line, Internet, 21 October 1997, available from http://cfa-www.harvard.edu/iau/lists/Closest.html. The asteroid's closest approach was 0.0046AU, approximately 1.77 times the distance to the Moon.

[9] American Institute of Aeronautics and Astronautics, "Responding to the Potential Threat of a Near-Earth-Object Impact," (Position Paper, AIAA, September 1995), 1.

[10] John C. Kunich, "Planetary Defense: The Legality of Global Survival," *The Air Force Law Review* 41, 1997, 119.

[11] Urias et al., sec 2.

[12] Ibid.

[13] Nici and Kaupa, 98.

Chapter 2

The Threat to Planet Earth

Toto, I have a feeling we're not in Kansas anymore.

—Dorothy, upon arriving in Oz

As children, many of us lay under the stars looking skyward to witness the brief but exciting streak of a meteor burning up in our protective atmosphere. Some of us still do. We naïvely envisioned that the flashes we saw resulted from the destruction of a large object. Scientists now know that the meteors constantly barraging Earth that give us a nightly display are generally no larger than a pea, weighing on average about one gram.[1] The larger objects we envisioned entering Earth's atmosphere do exist, but they don't always disappear in a fiery but harmless death. As we shall see, Earth has been impacted by large (tens to thousands of meters in diameter) objects in the past and will continue to experience this threat forever.

Celestial objects near enough to threaten Earth generally fall into two categories: 90% of potential Earth-impacting objects are Earth-crossing asteroids (ECAs) or short period comets. Collectively these are termed Near-Earth Objects (NEOs), all having orbits that intersect or closely approach Earth orbit. The remaining objects with potential to impact Earth are long period (greater than 20 years) comets.[2] NEOs make up the majority of the hazard potential because they are more numerous; however, long period comets cannot be detected as far in advance due to their orbital characteristics. The first

Earth-crossing asteroid was discovered only 60 years ago, and the first significant attention given to the problem posed by NEOs began just in the past 20 years. This attention began in 1980 when Luis Alvarez shocked the scientific community by proposing that a global dust cloud caused by the impact of a celestial body resulted in the extinction of the dinosaurs.[3]

In 1991, the House Committee on Science and Technology made a great leap forward recognizing the NEO threat in the NASA Authorization Bill: "The chances of the Earth being struck by a large asteroid are extremely small, but since the consequences of such a collision are extremely large, the Committee believes it is only prudent to assess the nature of the threat and prepare to deal with it. We have the technology to detect such asteroids and prevent their collision with the Earth."[4] The concepts, technology, and systems necessary for detection and collision prevention are collectively termed "planetary defense."

Congressional interest sadly was not a result of visionary leadership, but sparked by the no-warning near miss of asteroid 1989FC and the resulting far-sighted advocacy of the American Institute of Aeronautics and Astronautics.[5] The near-disaster had quickly transformed the NEO threat from the realm of science fiction to a serious, factual international issue.[6] Congress directed NASA to conduct two studies. The first was to "define a program for dramatically increasing the detection rate of Earth-orbit-crossing asteroids."[7]

NASA's report, submitted in 1992, recommended a program for detecting NEOs of approximately 1km diameter or larger, and is known as the Spaceguard survey. The second study was to "define systems and technologies to alter the orbits of such asteroids or to destroy them if they should pose a danger to life on Earth."[8] It determined many diverse technologies may exist in the future; however, for the present, only nuclear explosives appear viable for mitigating larger NEOs and those with short warning times.[9] In spite of the fact that Earth has been troubled by impacts from NEOs and comets for as long as the planet has existed, knowledge and recognition of a natural planetary threat is a very recent phenomenon.

Size vs Hazard

The Spaceguard survey developed three classifications of impactors based on the size and resulting kinetic energy of the impactor and the effects resulting from the impactor. The term "impactor" refers to Earth-crossing asteroids or comets, as the hazard resulting from each is primarily a function of their size and energy rather than origin. The first category includes the smallest impactors, which are generally disrupted in the atmosphere, with the atmosphere generally dissipating most of the energy. This impactor category primarily results in localized effects. In the second category, the body reaches the ground sufficiently intact to make a crater. Direct effects of the blast are primarily local; however, nitric oxide and dust can be carried for large distances. If the impact is in the ocean, a tsunami will result. The ocean impact is potentially more dangerous than a land impact because the tsunami can cause major destruction on any coastline of the impacted ocean, even if the destruction resulting from an equivalent land impact were relatively localized.[10] The third category covers a ground impact of sufficient size that the dust generated produces major short term climate changes and disastrous blast effects near the impact site.[11]

Category 1: 10-100m Diameter Impactors

Smaller impactors in this category are estimated to enter the atmosphere about once per decade, with approximate kinetic energy equivalent to 0.1 megatons of TNT. At the larger end of the scale, a 100m impactor has the energy equivalent of approximately 100 megatons, about the size of our largest thermonuclear weapons.[12] Spaceguard's estimate of the frequency of these occurrences is based primarily on the cratering record of the Moon and, to a lesser extent, Mars and Mercury. Earth's cratering record is not useful since there is no lingering evidence of atmospheric explosions, and ocean impacts as well as erosion mask the true record of Earth impacts. In contrast to this once-per-decade estimate, satellites recorded 136 atmospheric explosions in the megaton range from 1975-1992 due to the atmospheric entry of asteroids.[13] This may suggest that the once per decade estimate of the smaller impactors is conservative.

A 10-m object rarely reaches the ground to produce a crater, although some rare iron or stony-iron asteroids will do so. Generally the object is broken up during deceleration in the atmosphere and a shock wave is developed. Part of this shock wave is heat and light—a "meteoritic fireball" —with the remainder being a mechanical wave. Usually the burst is high enough above the ground that the mechanical wave has sufficiently dissipated when it reaches the ground that no damage results. With increasing size of the impactor, the energy in the wave increases to the point that both the mechanical and radiated energy cause terrestrial damage. This happened in the Tunguska impact, which resulted from an object approximately 60m in diameter exploding 8km (26,000 ft) above the ground. If such an event happened over a populated area, it would result in buildings being flattened within a 40km diameter area, with fires ignited near the center of the region.[14]

Category 2: 100m-1km Diameter Impactors

Impactors larger than 100m in diameter typically are not exploded in the atmosphere and survive to impact the ground, producing a crater. If the impact is on the ground, the crater will be approximately 3km in diameter. The ejecta, material displaced or "ejected" from the crater, will cover an area approximately 5 miles from the impact site. Thus, with the smaller category 2 objects, the area affected may actually be slightly smaller than that produced by a category 1 object due to the ground impact. In any case, the effects remain localized. If the impact is in the ocean, effects will be much more widespread. For example, Yabushita and Hatta analyzed the effect of a 200m object impacting in the Pacific Ocean. The expected height of the resulting ranged from 15 to 60m and "nearly all man-made buildings [on the Pacific's periphery] will be destroyed."[15] The expected frequency of occurrence for category 2 events is approximately once every 5,000 years.[16]

The effect of an impactor at the upper end of the category 2 size range, approaching 1km in diameter is less certain, chiefly because the energy associated with this size body *far exceeds what has been studied in nuclear war scenarios.* The Spaceguard survey extrapolated effects and estimated that the so-called "local effects" of the small category 2 bodies would expand to encompass far larger areas—whole states or nations—with hundreds of thousands of people killed, and hundreds of billion of dollars in damage. With increasing size, more global effects are produced with climatic cooling resulting from atmospheric dust similar to that evident in the largest volcanic eruptions.[17]

Scientists' ability to predict the effects of the impact of a comet in this size range are even less certain than predictions of asteroid impact because of differences in the composition of each. Asteroids are generally composed primarily of stone or metals,

11

while comets are largely composed of ice. Spaceguard predicted that a less than 1km comet could not survive passage through the atmosphere; therefore, atmospheric burst damage is expected.[18]

Category 3: 1km-5km diameter impactors

With an impacting body greater than 1km in diameter, serious global consequences are certain, although the mechanics of effects generated by material ejected into the atmosphere when the object impacts Earth (assuming a land impact or a very large body impacting the ocean) are not well understood and require further study. As a rule of thumb, the crater produced will be 10 to 15 times larger than the impactor—e.g., a 5km asteroid will produce a 50-75km crater. As a point of reference, the Washington, DC beltway is approximately 30km in diameter.

The primary hazard from a category 3 impactor is not the destruction associated with the impact crater or the associated blast effects, however, as it is with the smaller bodies. While the projectile and the impact area will be partially vaporized, and a horrific firestorm will be produced, the primary threat is from the dust and debris introduced into the atmosphere. This dust will lead to total darkness, which could last for many months. This is the effect generally described as nuclear winter—global temperatures could drop many tens of degrees; fresh water sources and possibly even the upper layer of the oceans would be acidified from nitrogen burning as a result of the firestorms. After the dust clears from the atmosphere, a greenhouse effect will be produced due to the excess water vapor released into the atmosphere—global warming. This effect could last for decades. If the impact were in the ocean, a tsunami several hundred meters high would result. Additionally, "hypercanes"—runaway hurricanes that inject large quantities of sea water into the atmosphere may be created which will also result in global climate changes.[19] Category 3 impacts are expected approximately every 300,000 years.[20]

An Earth impact by a celestial body of this size would be globally catastrophic: tens or hundreds of millions of deaths, massive starvation, extinction of species—perhaps our own—and possibly an end to civilization as we know it. The size threshold for an impactor to cause massive global effects is uncertain, but there are a few data points demonstrating that large impactors have clear global impacts. Be forewarned, there is no good news. From analysis of lunar cratering, scientists expect a 10km body to impact Earth every 50 million years. This is the size impact the Alvarez theory says caused the extinction of the dinosaurs approximately 65 million years ago. This impact, near the Yucatan peninsula in Mexico resulted in 100 trillion tons of debris being lofted into the atmosphere.[21] This amount of material would be *several thousand* times more effective at blocking sunlight than the "nuclear winter" scenario, which assumes 10,000 nuclear explosions.[22] While much of the scientific community was skeptical of Alvarez when his theory was first presented in 1980, there is now general consensus that a NEO impact was the primary cause of the demise of the dinosaurs and the end of the Cretaceous period.[23]

Scientists at the University of Oregon have found evidence of an asteroid impact 250 million years ago that hit in the South Pacific Ocean south of Australia producing a crater 300 miles in diameter. They believe this was the cause of over 90 percent of all life forms disappearing from Earth, including entire classes of plants, reptiles, and shellfish.[24] At least 6 mass extinctions have been linked to asteroid impacts,[25] with Raup and Sepkoski having determined a periodicity of mass extinctions every 26 million years. Two separate groups have found known craters which correlate with this periodicity, within the accuracy of the technology used to date the craters.[26] Frank has theorized the existence of a large, as yet undiscovered planet which dislodges comets from the Oort disk to account for this periodicity.[27]

Unique Aspects of Comets

As described above, the most serious hazards from an object impacting Earth are from large Earth-crossing asteroids, because their composition causes them to survive their flight through the atmosphere to impact land or ocean. Comets bring their own problems, however. The Spaceguard survey grouped short period comets with Earth-crossing asteroids, but separated out long period comets primarily because due to their orbital characteristics, we will be unable to detect them until they are relatively near Earth. Although the number of long period comets is estimated to be only 5 to 10 percent of the Earth-crossing asteroids, they are estimated to comprise 25 percent of the Earth impact hazard.[28] Long period comets will pass through the solar system at most once during a period when NEOs are being surveyed (10-25 years). Current estimates are 180 long period comets of greater than 1km diameter cross Earth's orbit annually. As they are generally not detectable until they are inside Jupiter's orbit, we should expect warning times of about 1 year. To achieve even this short warning time would require a robust detection program which does not now exist.[29]

Risk

The intent of the above discussion was to clearly explain that a range of potential Earth hazards do exist; that is, given that an Earth impact occurs, we know we can expect varying degrees of disastrous consequences. All of these consequences are beyond our experience and many of them are beyond our imagination or comprehension. But the severity of the hazard is not the only element in a discussion of risk. The other element is the probability that such a hazard will occur, and our ability to comprehend and act on that probability.

Probability of Occurrence

Perhaps it's best to begin the discussion of probability with what is certain (probability 1.0) and what is not (probability 0.0). We know *for sure* (probability 1.0) that in the future, Earth will collide with asteroids and comets. The recent near misses and atmospheric explosions that were used to introduce this paper demonstrate that collision with these objects is still a threat. Astronomer Ralph Baldwin wrote in 1949

> . . . since the Moon has always been the companion of the Earth, the history of the former is only a paraphrase of the history of the latter. The study of the Moon thus gives us a mirror throughout all time with which to study our own Earth. [Yet] the vista opened up . . . contains a disturbing factor. There is no assurance that these meteoritic impacts have all been restricted to the past. Indeed we have positive evidence that meteorites or asteroids of the requisite size still abound in space and occasionally come close to the Earth. The explosion which formed the crater Tycho on the Moon left us an interesting [structure] to study. A similar occurrence anywhere on the Earth would be a horrifying thing, almost inconceivable in its monstrosity.[30]

Baldwin certainly was convinced, as are most scientists, that the past is the best predictor of the future. Looking at a plot of Earth's orbit compared to the hundred largest ECAs (Figure 4) visually demonstrates the fact that we live in a cosmic shooting gallery; the question of an asteroid or comet impacting Earth is not a question of *if*, but solely a question of *when*. As for what we are sure will not occur, the answer is very little. One might think that we could rule out a threat from known asteroids whose orbits do not intersect Earth's orbit. In truth, asteroid orbits are subject to change over time. For example, Michel *et al.* studied the evolution of the 22 km diameter asteroid 433 Eros, whose orbit does not now intersect Earth's.[31] Their simulation of the "chaotic nature" of asteroid orbits, including orbital perturbations brought about by the gravitational effects of other bodies, notably planets, demonstrated that 433 Eros' orbit could over time evolve such that it may become an Earth-crossing asteroid.[32]

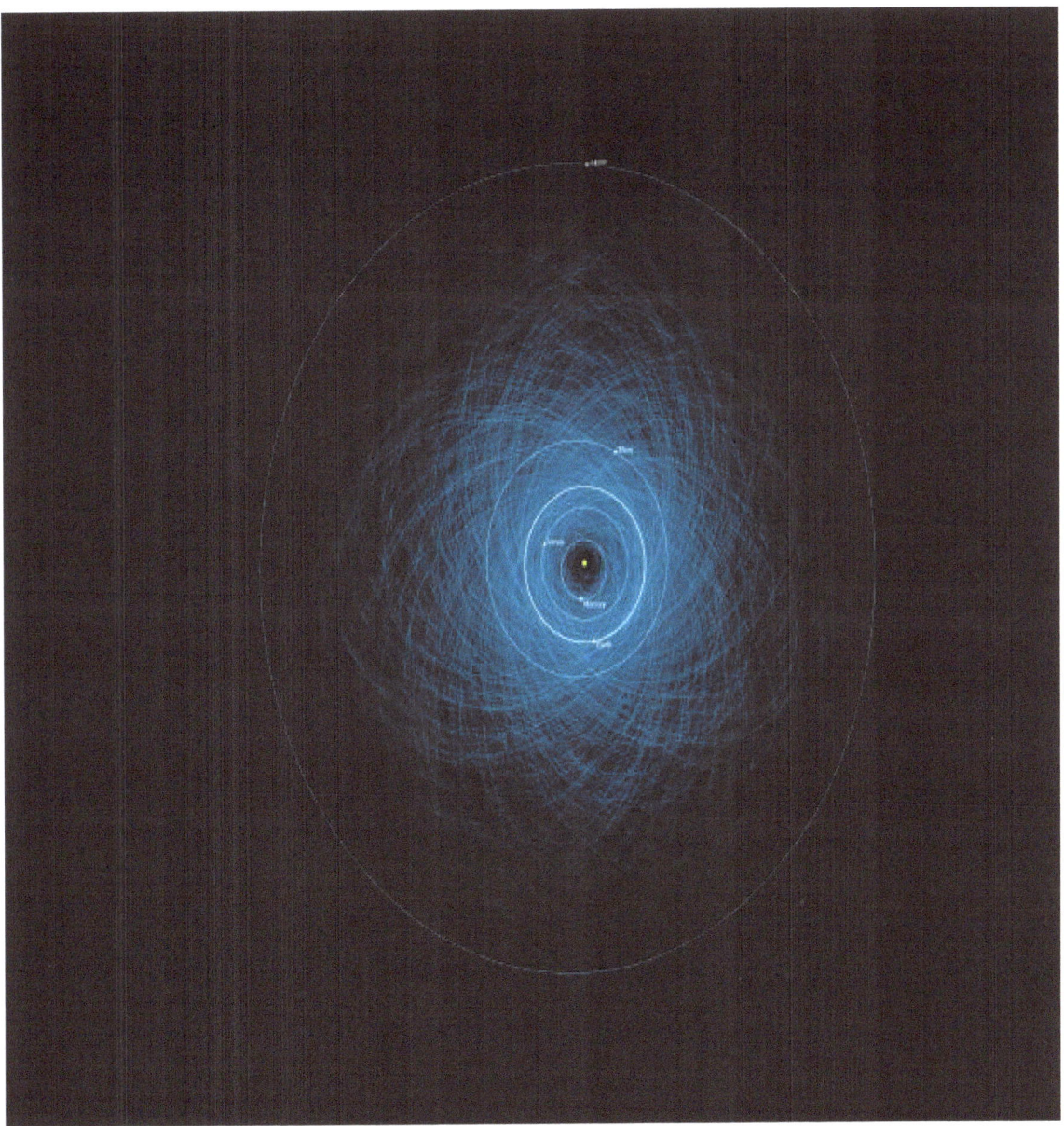

Figure 4. Earth Orbit Vs 100 Largest Earth-Crossing Asteroids

The "when" aspect is what makes the NEO threat difficult to grapple with. Having become aware of the threat, our first question must be do we know of any NEOs that will impact Earth at some definite, known time in the future? The short answer is no—but. At the time of the Spaceguard study in 1991, 128 Earth-crossing asteroids had been identified. None are a current hazard, meaning there is no definitive date when they will collide with Earth, although the scientists estimated that eventually 20 to 40 percent of these will do so.[34] That no definitive near term threat exists appears comforting, except that scientists also estimate that *only 1 percent* of the Earth-crossing asteroids of sufficient size to threaten Earth have been found.[35]

Progress in identifying and cataloging NEOs is agonizingly slow since there is no formal program for conducting this work. Fewer than a dozen people worldwide are currently working the problem.[36] As of 6 January 1998, the NEO data base had only increased by 25 objects to 153—clearly unremarkable progress since 1991.[37] So small is the fraction of space being monitored for NEOs that the most probable warning time for an object in the 1km class is zero.[38] With that as a framework, let's consider some probabilities. Based on the impact density of lunar craters, we would expect a Tunguska class event roughly every century, and an impact the size of the one which caused the extinction of the dinosaurs every 25 million years.[39] Ready to resupply the bomb shelter your father built in the late 1950s? Try these numbers. Your risk of experiencing a civilization-ending impact is[40]

- About 2,000 times greater than your risk of dying from exposure to trichlorethylene[41] at the limit established by the Environmental Protection Agency
- About 300 times greater than your risk of dying from botulism poisoning
- About 100 times greater than your chance of dying from a fireworks-related accident
- About 10 times the chance you will die in a tornado
- About one third the risk of death by a firearm accident

- About 1/30th the chance you will be murdered
- About 1/60th the chance you will die in an automobile accident

Bell, *et al* researched and analyzed the varying probabilities published for an Earth-impacting object and found much of the data to be outdated or otherwise flawed, with the data produced by Chapman and Morrison and presented above assessed as the most credible.[42]

In the short history of planetary defense discussions, the "giggle factor" is most often centered on the probability of the event, generally being dismissed as being ultra-unlikely, without considering the consequence of the hazard. While the hazard is extreme, ranging to the destruction of all that we know, development of a believable and understandable probability of occurrence is still problematic at best. This is an excellent example of the old adage "liars figure and figures lie." Opposing points could be argued, likely accompanied by statistics from a source associated with the hazards from NEOs and long period comets. The issue boils down to this: *when* will Earth be impacted by a NEO, and will we, on a well-informed basis, *accept the risk* that it won't happen soon enough that we need be concerned with it now.

Risk Acceptance

Most people tend to view risk levels for most activities as unacceptably high, suggesting that they are dissatisfied with the way market forces or regulatory restrictions have balanced risk and benefit.[43] Even so, the public is generally unwilling to accept the notion of risk associated with an activity or event without some level of credibility in the risk assessment. Both the public and government agencies tend to ignore anything that is not a "proven" hazard, thereby assigning it a zero risk.[44] This suggests that we might expect the risk associated with a NEO impact to be either overstated or understated.

In fact, this is exactly the case—previous experience with other natural and manmade risks shows that there are a number of factors that would tend to predict both high and low public concern for the NEO threat. We would expect the public to view the threat of an impact as a high risk, sufficiently high to warrant action to be taken because:

- "The risk is demonstrable (it happened to the dinosaurs) and is endorsed by credible scientists.
- The potential consequences of large impacts are uniquely catastrophic and are qualitatively different from other natural hazards.
- The probabilities of catastrophic impacts, while small, are not trivial. Considerable public funds are already being spent to deal with risk of even lower probability, such as death or injury from tornadoes or terrorist attacks.
- Unless action is taken, the risk is unknown and uncontrollable."[45]

Conversely, there are also factors that would suggest from previous experience that this threat would be viewed with little concern.

- "Natural hazards such as impacts tend to be less frightening than technological hazards. People perceive nature as benign and react rather apathetically to the threat from natural hazards. Personal experience of a natural disaster is usually necessary to motivate action to reduce future risks.
- Probabilities are typically more important than consequences in triggering protective actions; hence the impact probabilities may be too low and the risk apparently too remote in time to trigger concern, in spite of their high consequences.
- People are often insensitive to very large losses of life. We will expend great effort to save an individual life, but in a context of impersonal numbers or statistics, the lives of individuals lose meaning. A threat that puts 100 people at risk is likely to be seen as quite serious, but we will probably respond identically to a hazard that threatens 2,200 people and one that threatens 2,300.
- People tend to prefer 100% insurance against a threat. If impact defense systems cannot provide 100% protection, they may be undervalued."[46]

There is no clear-cut answer, then, based on general experience with other hazards; the potential of a NEO impacting Earth contains elements that would predict both high and low public concern.

"Experts" tend to view risk as synonymous with expected mortality rates[47]—a rather cold, antiseptic approach, and no good public barometer reacting to the NEO threat has yet materialized. While public knowledge of the NEO threat has existed for several years based on occasional brief coverage in the national media, the public is not well informed and no large-scale opinion survey has yet been conducted. A small-sample survey was conducted about 5 years ago, with a group of approximately 200 people who in previous surveys had been shown to respond similarly to larger, more broadly based demographic groups. After reading several media articles dealing with the asteroid impact threat, this group ranked the threat 14th in a list of 24 hazards assessed; however, they had considerable difficulty in dealing with the immediacy of the threat due to the low probability figures. For example, only 6 percent of the survey participants believed that an impact *could* occur in the next 50 years. Additionally, one-third assessed a statement concerning a civilization-threatening impact as "not believable."[48]

Public acceptance of the reality of potential impacts has not been assisted by Hollywood; in fact, the movie industry may have contributed to many "not believable" attitudes. Productions dealing with asteroid impacts to date have been "disaster" films of the genre begun by *The Towering Inferno*. Unfortunately, many view these films as fictionalized, pure entertainment with little if any basis in reality. Is an attack by asteroids from space really any different than an attack by menacing aliens in *Independence Day*?

Two more impact films are nearing release: *Deep Impact*, dealing with a threatening comet is planned for release in May 1998, and July 1998's *Armageddon* involves an asteroid "the size of Texas."[49] Might we expect each film's star to defeat these errant bodies just in the nick of time? And will it be done with Yankee bravado based on a plan concocted over a few days or hours; or with a program born of many years' careful

20

planning, design, testing and construction to place it in readiness? While the bravado is a virtual certainty and will sell far more theater tickets, it will result in even greater public difficulty in dealing with the true risk of an impact.

Difficulty in dealing with low probabilities of occurrence correlates well with Starr and Whipple's view that perception of risk is nonlinear. For very low probability events, the intuitive perception underestimates the "true" or quantitative probability, even to the point of estimating it to be zero. As the quantitative probability increases, the intuitive estimate overestimates the probability.[50] Thus, for a low probability, high consequence event such as a NEO impact, a conflict in the evaluation of the hazard exists.

Sociologists tell us that the public will accept risks over which they have some control—driving a car, skiing, mountain climbing, and the like. Presumably the individual would view the risks taken as a predominantly random process; his measure of control diminishes those risks. On the other hand, they are far less tolerant of risks over which they have no control—airline travel (the pilot has the control), environmental hazards, etc. Chauncy Starr, a risk assessment pioneer, asserts the public will accept voluntary risks 1000 times greater than involuntary risks.[51] Risk acceptance is also influenced by the conditional probability of survival given an accident. This is another factor in some people's greater fear of air travel over automobiles even though air travel is statistically safer.[52] Some sociologists have estimated that a risk of death of 1 in 1 million is the public's threshold of concern.[53] Each of these attributes of risk acceptance suggests that the NEO threat should be one of public concern. It is clearly an involuntary risk—we as individuals have no influence over the orbits of comets and asteroids. Any person in the area affected by an impact (regional to global depending on the size of the impactor) has a low probability of survival given that the impact occurs. Finally, the risk of death far exceeds the 1 in 1 million criterion, as the risk of a globally catastrophic event over a lifetime is estimated to be approximately 1 in 10,000.[54]

Slovic has suggested an analytic representation or risk based on two factors, "dread risk" and "unknown risk." Dread risk is defined at the high end by perceived lack of control over the risk activity, dread, catastrophic potential (as opposed to effects limited to an individual), fatal consequences, and inequitable distribution of risks and benefits. High unknown risk is characterized by hazards judged to be unobservable, unknown to those at risk, new, and delayed in their manifestation of harm.[55]

The risk associated with an impacting asteroid or comet would score highly on both these scales. In terms of the elements of dread risk, each is applicable to the NEO threat except for inequitable distribution of risks. Since no point on Earth and no individual or population is more or less likely to be affected by a future impact, the distribution of risk is the same or equitable for all. Examining unknown risk, the impactor is not always observable to those at risk; when it is observable, we can observe it only immediately before impact. Those at risk are unaware of the danger beforehand, and although this threat is not "new" in an astronomical sense, it is relatively new in that we have only recently understood it to be a continuing threat of potentially catastrophic proportions. The NEO threat is not consistent with the last element of unknown risk, delayed manifestation of harm, because the effects resulting from an impact begin immediately.

"Research has shown that lay people's risk perceptions and attitudes are closely related to the position of a hazard within this type of factor space. Most important is the . . . factor "dread risk." The higher a hazard's score on this factor . . ., the higher its

22

perceived risk, the more people want to see its current risks reduced, and the more they want to see strict regulation employed to achieve the desired reduction in risk."[56]

Other measures of risk acceptance have been problematic in their ability to provide a clear estimate of the risk associated with the low probability, high consequence event of a celestial object colliding with Earth. Slovic's analytic representation, however, provides a reasoned basis for concluding that a reasonably informed public will likely conclude that action should be taken to mitigate the NEO threat.

A Signal

"The informativeness or 'signal potential' of an event, and thus its potential social impact, appears to be systematically related to the characteristics of the hazard and the location of the event within the [dread and unknown risk parameters] described earlier. An accident that takes many lives may produce relatively little social disturbance (beyond that experienced by the victims' families and friends) if it occurs as part of a familiar and well-understood system (such as a train wreck). However, a small accident in an unfamiliar system (or one perceived as poorly understood), such as a nuclear reactor or a recombinant DNA laboratory, may have immense social consequences if it is perceived as a harbinger of further and possibly catastrophic mishaps."[57] Additionally, media attention and elevated costs of a mishap create higher signal potential.[58]

The effect the no-warning passage of asteroid 1989FC had on Congress has already been discussed. It produced some elegant rhetoric that Congress was prepared to "deal with" the threat of potential impact with NEOs. Action was sharply limited as today funding for "dealing with" the NEO threat is nil, and more importantly, no policy exists in this area. Suppose 1989FC had piqued the interest of the media beyond merely reporting that its close passage occurred, or the logical analogy to Earth had been more forcefully drawn during the 1994 impact of Shoemaker-Levy 9 on Jupiter. What if CNN, network television news, and the print media had presented daily predictions of the damage if 1989FC or a larger body had hit Earth? Imagine the onslaught of investigative reports citing the inability of the scientific or military communities to detect the object as well as thousands of others, and decrying the absolute dearth of capability or inclination to respond if necessary.

What if today, Congress were holding hearings on why NASA, the Department of Defense, and the Federal Emergency Management Agency have done nothing to address this catastrophic risk, rather than holding hearings on campaign finance improprieties? Now *that's* a signal. Or is it an outlandish scenario? Suppose an authoritative source were to claim publicly that there is a very low (on the order of an asteroid impact) probability that any of the nuclear warheads on Earth could spontaneously detonate—in other words, all the safeguards fail. Would the public reaction be to immediately make whatever investment is required to mitigate the risk, or to dismiss it as such a remote possibility that we need not concern ourselves with it since none have spontaneously detonated since 1945? It's difficult to imagine that we would not immediately make the necessary investment.

Why is this "signal potential" important in the risk assessment equation? The analytical methods used by regulators of public risk assessment can be easily overruled through the political process. A good example was the repeal of the once-required seat belt starter interlock that was installed as an automobile safety device.[59] Though it had great promise to reduce the risk of fatality or serious injury in an automobile crash through the forced use of seatbelts, it died a rapid political death in the name of consumer convenience.

Unnatural Power

The preceding discussion approaches risk acceptance from a logical perspective. Humans by nature, however, are not always logical in their analysis of problems or in their decisions. Science journalist Oliver Morton has considered the reasons for widespread denial that we humans have a responsibility to take action to prevent catastrophic impacts. He suggests that, at a governmental level, denial exists because "political changes need constituencies, and 'people who will be harmed by an impact' simply do not make up an identifiable constituency."[60] More interesting, however, is Morton's conclusion that individual denial is deep rooted: people are innately disturbed by the potential for man to wield great power over his universe.

> Perhaps people do not want to see themselves connected to the universe in this sort of way [diverting an incoming comet or asteroid]. The geologists who for years resisted the impact explanation for the dinosaurs' death simply didn't want asteroids to play as big a role in the history of the Earth as, say, the wanderings of one of its own tectonic plates. Tough: they do. Humans and the Earth they live on are linked to the universe in all sorts of strange, indirect, unsettling ways—and, worse yet, humanity now has the power to change these connections. We can empty seas and denude vast forests, we can warm an entire planet and now, given just a little warning, we can push aside flying mountains. It's genuinely frightening to contemplate such power, especially when you realise how poorly decisions about using it are made or not made. Better to deny the risk of asteroid impacts than to accept the fact that humans can redirect the stars in their courses. It's a delusion; in this case a slightly dangerous one—but you can understand it.[61]

What Does It Mean?

"Perhaps the greatest intellectual challenge in dealing with this threat is the extraordinarily low annual likelihood coupled with its incomparably dire consequence."[62]

It's important to remember that a globally catastrophic event would be unique in recent human history—one hasn't happened since we started keeping records, save perhaps the Great Flood of *Genesis*. Our history and our priorities are typically driven by something that happens to us or to someone we know or know about—that is, relevant history. Our priorities will be shaped by public perception based on the factual presentation of understandable data and by the relentless media pursuit of an issue. Studies support the notion that people are willing to tolerate higher risks for the promise of increased benefit.[63] The logical converse of this is that people are unlikely to accept risks for which there is no benefit, such as an asteroid or comet impact.

The consequence of a globally catastrophic impact is what sets it apart from everything we can envision except full scale global thermonuclear war, assuming that we were even capable of fully appreciating those impacts. Currently we say that the probability of an Earth impact is extremely small based on the cratering history of the Moon, Mars, and Mercury. Further, we can say that none of the *known* Earth crossing asteroids are projected to impact Earth at any specific date in the future, but a large percentage of them are likely to impact Earth sometime. But we also know that we have identified only a *very small* fraction of the total population of Earth crossing asteroids, and we are unable to identify long period comets until they are quite close to Earth. Therefore, we all *must* conclude that one day in our future—as early as tomorrow and as many as tens or perhaps hundreds of years from now—Earth *will with absolute certainty*, meaning with probability 1.0, be struck by an Earth crossing asteroid or comet, resulting in catastrophic destruction.

While this threat has not been hidden from the public, it has been received little serious scrutiny in the media and has been fictionalized to the point of surrealism by Hollywood. It has not risen to a sufficient level of consciousness in the public eye, or at any level other than in a small sector of the scientific community, to generate sufficient concern that a concerted effort can be made to evaluate the threat and decide what action to take. Perhaps, as Morton suggested, we are so concerned about our ability to control the universe that, to our detriment, we neglect the universe's ability to destroy us. We must now evaluate just how sobering the prospect of an asteroid or comet impact with Earth is, deciding whether or not to implement solutions to the problem of defending our planet.

We must be judicious in how we go about risk assessment on this subject, however, because research "indicates that disagreements about risk should not be expected to evaporate in the presence of evidence. Strong initial views are resistant to change because they influence the way that subsequent information is interpreted. New evidence appears reliable and informative if it is consistent with one's initial beliefs; contrary evidence tends to be dismissed as unreliable, erroneous, or unrepresentative."[64] We must assure that we are guided by the facts, not confused by them.

Notes

[1] Space Science Division, NASA Ames Research Center, *Spaceguard Survey: Report of the NASA International Near-Earth-Object Detection Workshop*, report submitted to Congress, 25 January 1992, n.p.; on-line, Internet, 16 September 1997, available from http://ccf.arc.nasa.gov/sst/spaceguard.html.

[2] Ibid., sec 1.

[3] Ibid.

[4] "Congressional Statements on the Impact Hazard," *NASA Ames Space Science Division*, 15 October 1996, n.p.; on-line, Internet, 16 September 1997, available from http://ccf.arc.nasa.gov/sst/c_statements.html.

[5] The media covered this incident as an event. It was generally not portrayed as an

issue for government action.

[6] George J. Friedman, "Risk Management Applied to Planetary Defense," *IEEE Transactions on Aerospace and Electronic Systems* 33, no. 2 (April 1997): 722-723.

[7] Congressional Statements on the Impact Hazard.

[8] Ibid.

[9] Gregory H. Canavan, Johndale C. Solem, and John D. Rather, "Near-Earth Object Interception Workshop," in *Hazards Due to Comets and Asteroids*, ed. Tom Gehrels (Tucson, Ariz.: The University of Arizona Press, 1994), 122-123.

[10] S. Yabushita and N. Hatta, "On the Possible Hazard on the Major Cities Caused by Asteroid Impact in the Pacific Ocean," *Earth, Moon and Planets* 65 (1994): 7.

[11] Spaceguard, sec 2.

[12] Ibid.

[13] Nici and Kaupa, 95-96.

[14] Spaceguard, sec 2.

[15] Yabushita and Hatta, 12.

[16] Spaceguard, sec 2.

[17] Ibid.

[18] Ibid.

[19] Nici and Kaupa, 96.

[20] Spaceguard, sec 2.

[21] Chapman and Morrison, 86-93.

[22] Ibid., 123.

[23] AIAA.

[24] Associated Press, *Scientists say asteroid slammed into Earth 250 million years ago,* Colorado Springs Gazette Telegraph, October 30, 1996 at A7, in Kunich, 124n23.

[25] Kunich, 124.

[26] Chapman and Morrison, 100-104.

[27] Louis A. Frank with Patrick Huyghe, *The Big Splash* (New York: Birch Lane Press, 1990), 168-175. Frank's proposed "Dark Planet" would lie in a highly elliptic orbit inclined to the orbital plane of the known planets. With an orbital period of approximately 500,000 years, orbital disturbances would result in it passing through the inner edge of the Oort disk every 26 million years. During this passage, where, along with the Oort cloud most comets are thought to originate "tens or hundreds" of large comets are propelled toward Earth, resulting in a high incidence of comets transiting the inner solar system for approximately 10,000 years. Frank also uses this planet to explain his belief that a constant barrage of "small comets" which burn up daily in Earth's atmosphere exists. To date, the scientific community has largely dismissed the existence of "small comets;" however Alvarez's proposal that a comet or asteroid impact resulted in the extinction of the dinosaurs was not initially accepted either.

[28] Spaceguard, sec 3.

[29] Ibid., sec 5.

[30] Chapman and Morrison, 46.

[31] 433 Eros has a perihelion of 1.13 astronomical units (AU).

[32] P. Michel, P. Farinella, and C. Froeschlé, "The Orbital Evolution of the Asteroid

Notes

Eros and Implications for Collision with Earth," *Nature* 380, no. 6576 (25 April 1996): 689-691.

[34] Ibid., sec 3.

[35] Bell, Bender, and Casey, 19.

[36] Spaceguard, sec 4.

[37] "Spacewatch Discoveries," *Spacewatch Home* Page, n.d., n.p.; on-line, Internet, 23 February 1998, available from http://xlr8.lpl.arizona.edu.spacewatch/ discoveries2.html.

[38] David Morrison, "Is the Sky Falling?" *Skeptical Inquirer* 21, no. 3 (May/June 1997): 23.

[39] Nici and Kaupa, 96.

[40] Chapman and Morrison, 283.

[41] Trichlorethylene is a carcinogen now being strictly regulated by the US Government.

[42] Bell, Bender, and Carey, 82-89.

[43] Paul Slovic, "Perception of Risk," *Science* 236, no. 4799 (17 April 1987): 283.

[44] Richard Wilson and E.A.C. Crouch, "Risk Assessment and Comparisons: An Introduction," *Science* 236, no. 4799 (17 April 1987): 268.

[45] David Morrison, Clark R. Chapman, and Paul Slovic, "The Impact Hazard," in *Hazards Due to Comets and Asteroids*, ed. Tom Gehrels (Tucson, Ariz.: The University of Arizona Press, 1994), 82.

[46] Ibid.

[47] Slovic, 283.

[48] Morrison, Chapman, and Slovic, 83-84.

[49] Based on January 1998 trailers for both films. Texas' maximum north-south dimension is 1,289km, and its maximum east-west dimension is 1,244km.

[50] Chauncey Starr and Chris Whipple, "Risks of Risk Decisions," *Science* 208, no. 4448 (6 June 1980): 1116-1117.

[51] Chapman and Morrison, 281.

[52] Starr and Whipple, 1116.

[53] Bell, Bender, and Casey, 89-90.

[54] Spaceguard, Sec 2.

[55] Slovic, 282-283.

[56] Ibid., 283.

[57] Ibid., 284.

[58] Ibid.

[59] Starr and Whipple, 1116.

[60] NEO News, NASA Ames Research Center, 10 February 1998. While this source is generally distributed only within the NEO-interest community, it should eventually appear on NASA's impact web site at http://arc.nasa.gov/sst/news/hot.html.

[61] Ibid.

[62] Friedman, 721.

[63] Slovic, 283.

[64] Ibid., 281.

Chapter 3

Potential Solutions

If some day in the future we discover well in advance that an asteroid that is big enough to cause a mass extinction is going to hit the Earth, and then we alter the course of that asteroid so that it does not hit us, it will be one of the most important accomplishments in all of human history.

—U.S. Rep. George E. Brown, Jr.

Given that NEOs and long period comets pose a threat to Earth, what capabilities do we now have to defend ourselves against the natural threat? There are three aspects of a planetary defense system.

1 Surveillance of space to identify and track potential threats
2 Control of the surveillance function to direct efficient operation among various search sites (ground or space) and cataloging of all data
3 Mitigation of the threat once it is identified. Mitigation ranges from deflection or destruction of the object to evacuation of an impact zone for small objects with minimum or no warning *if* an impact zone could be accurately predicted.

At the current rate of NEO detection, it will take more than a century to catalog even the larger objects.[1] Relying on this rate of progress for adequate warning of a potential impact is marginally better than reliance on blind luck, fortune-tellers, or magic potions. Our lack of a surveillance program is not a fatal flaw, however, because we currently have exactly zero capability to mitigate a known future impactor, and no agency is assigned any responsibility or has any current or future programmed funds to develop or implement a mitigation system. In other words, even if we discover we have a problem, we can't do anything about it, and it's no one's job to fix it! Let's look at some potential solutions, what they may cost, and what legal barriers to implementation may exist.

Surveillance and Control

The objective of a surveillance system is to locate objects with periodic approaches to Earth, calculate their orbits, and determine which objects, if any, will enter the capture cross section of Earth.[2] Scientists *expect* that when a potential impacting NEO is located, we will have at least several decades warning time.[3] The technology required to accomplish adequate surveillance is not exotic. Obviously, the first priority is to locate Category 3 objects first. The bodies we seek can be found at the inner edge of the main asteroid belt, about 200 million kilometers distant. At that distance, they cannot be detected with existing radars, but modest optical telescopes will do the job. The Spaceguard survey determined that a 1km object equates to a stellar magnitude 22 target, which can be identified and tracked with a ground-based 2m aperture telescope.[4] Space-based telescopes could also be used, but the additional expense is probably unjustified, at least until significant progress is made in identifying the larger NEOs.

Spaceguard proposed a network of 6 telescopes, each with a 2-3m aperture. Each would be a new instrument so as not to compete for sky time with existing telescopes, but each would be sited at an existing observatory to minimize infrastructure costs.[5] To achieve adequate sky coverage, the telescopes must be widely separated in both the northern and southern hemispheres, dictating international cooperation. The Spaceguard survey estimated that if this network were activated, greater than 90% of NEOs greater than 1km in diameter would be identified within 20 years; incoming long period comets could be detected with approximately a 1 year lead time prior to reaching Earth's orbit.[6] This is a relatively noncomplex approach and indeed may be adequate.

Bell, *et al* believe, however that is absolutely necessary to find the larger objects, but we should not ignore the smaller, more abundant objects that can still cause disastrous damage and loss of life. Their proposal is the surveillance system should be able to detect objects in the 50m class; or if technologically infeasible, the 200m class as this size impactor generally survives atmospheric reentry to cause an impact crater.[7] It is estimated that there are 4-10 million asteroids with diameters 50m or greater. In fact, 6,000 asteroids greater than 100m in diameter pass within 5 million miles of Earth annually.[8]

Bell *et al* propose two additions to the Spaceguard recommendations. First is the addition of an infrared search component to examine areas that are blinded by having look angles in close proximity to the sun. They believe that an Explorer class spacecraft such as a modification to the Wide-Field-of-View IR Explorer (WIRE), or alternately hosting an infrared telescope on the forthcoming Space Station is the most economical solution.[9] Additionally, they propose upgrades to the Air Force's Ground-based Electro-Optical Deep Space Surveillance System (GEODSS) and use of liquid mirror telescopes developed by NASA to augment or more likely substitute for the optical network proposed by Spaceguard.[10] Regardless of the specific system chosen to perform surveillance tasks, the job is not a finite one. A survey that must detect long period comets can never be complete, since new comets are constantly entering the inner solar system.[11]

A control function is needed to coordinate the efforts of the worldwide (or orbital) sensors conducting surveillance to ensure efficient, nonduplicative data collection, catalog and store the data collected, and perform data analysis. In other words, the people and computers that comprise the control center are "in charge" of making sure the entire surveillance program works. The requirement for a control center is evident from the fragmentary nature of the current survey program, which has led to NEOs being discovered and subsequently lost due to insufficient tracking before a reasonably precise orbit could be calculated.[12]

The responsible agency to accomplish this function must be chosen carefully. There are likely three organizations with the potential capability to be the NEO surveillance control center if the necessary processing and communication upgrades are implemented. First is the International Astronomical Union's Central Bureau for Astronomical Telegrams and Minor Planet Center at Cambridge, Massachusetts. Also in the running are the NASA Jet Propulsion Laboratory and the military's US Space Command. The Minor Planet Center accomplishes the majority of this function now with data collected through the Spacewatch network. Additionally, they have significantly more NEO search experience than either of the other organizations and coordinate extensively internationally. US Space Command currently performs a similar function tracking orbital man-made debris.[13]

Mitigation

The objective of a mitigation system is to deflect an impactor such that its trajectory is sufficiently altered to avoid entering Earth's capture cross section; alternately, to destroy or fragment the object sufficiently that it is no longer a threat. Several factors affect the design and effectiveness of any mitigation system: the distance at which the object will be engaged and the shape, size, composition, and inherent motion (e.g., spin) of the object. Distance is important because the required deflection angle or change in velocity is reduced at greater distances. Engagement distance will be driven primarily by the warning time available and the speed with which the mitigation system can act on the object. Warning time is a function of the orbital characteristics of the impactor and the capability of the surveillance system; action time is a function of the mitigation system. For example, a directed energy system can act quickly, while a booster with an explosive device must be readied, launched, and travel the distance to the object. Detection is generally not a function of the object's composition, but is primarily driven by size. The remaining physical characteristics are important because they affect the suitability of various mitigation techniques.

Our present knowledge of the physical characteristics of asteroids and comets is limited. Knowledge of asteroids is derived primarily from comparison of spectra from bodies in the main asteroid belt and investigation of the properties of meteorite fragments found on Earth. One asteroid—951 Gaspra—has been studied by a spacecraft; several flyby spacecraft studied Halley's Comet in 1986.[14] In the absence of effective active mitigation, improved knowledge of the physical characteristics of asteroids and comets is needed to react to potential impacts. For example, the Tunguska and Arizona impactors were of roughly the same size, yet one exploded in the atmosphere and one impacted the ground.[15] We know that small stony or carbonaceous asteroids are more likely to explode in the atmosphere than an asteroid composed largely of iron, although their potential for disruption is strongly a function of their relative velocity entering the atmosphere.[16] Better knowledge may help understand the interaction of these bodies with the atmosphere.

Numerous concepts have been suggested for potential mitigation systems, ranging from the obvious to those reminiscent of *Star Trek*. Among these concepts are rockets with nuclear, chemical, or antimatter explosive devices; rocket propulsion systems (to propel the object into a non-threatening trajectory); kinetic energy systems; high energy lasers; microwave energy systems; mass drivers/reaction engines; solar sails; solar collectors; biological/chemical/mechanical NEO "eaters;" supermagnetic field generators, force shields, tractor beams, and gravity manipulation.[17] For all who lack the theory underlying these concepts and a technology forecasting background, the giggle factor in some of these mitigation concepts would likely dwarf the giggle factor accompanying the uninformed notion of planetary defense itself. Without engaging a discussion of the scientific merits of these concepts, suffice it to say that for the relatively near future, the only practical mitigation concepts are use of a kinetic energy impact (non-explosive) to deflect a threatening object, or a nuclear device to deflect, fragment, or destroy it.

Scientists believe the technology for a nuclear solution exists to deflect an incoming object or to fragment it into sufficiently small pieces (less than 10m diameter).[18] Ahrens and Harris determined the velocity change necessary to be imparted to a potential impacting asteroid to divert it from collision with Earth. They determined that with a 10 year lead time, the velocity increment is only 0.1 meters per second. This change in velocity is most effective when applied at the asteroid's perihelion.[19] Decreased lead time obviously will lead to increased delta velocity requirements. With this required velocity data, Ahrens and Harris studied in detail 3 candidate mechanisms for deflection: a large mass striking the asteroid at high velocity (kinetic energy), a propulsion system attached to the asteroid (mass driver), and nuclear explosions which cause deflection either through the effects of radiation, or the impulse created due to cratering. Additionally, they studied fragmentation of the asteroid by a subsurface nuclear explosion.[20]

They concluded that a kinetic energy system would be effective for deflecting a potential impactor of approximately 100m in diameter. For larger bodies, however, the increased gravity of the object would reduce the mass of material escaping the asteroid's gravity as a result of the impact such that the asteroid's orbit would be insufficiently perturbed to avert impact with Earth. Accordingly, a kinetic energy system to mitigate objects larger than 100m is judged impractical. A propulsion (mass driver) system would be required to operate for approximately 30 years to deflect a 1km asteroid; this technique is judged inefficient compared to deflection by nuclear explosion.

Ahrens and Harris analyzed two nuclear deflection techniques—deflection by radiation (explosion above the surface) and deflection by surface explosion. In the first case, detonation of the nuclear device at a precise distance above the surface of the asteroid causes approximately 30 percent of its surface area to be irradiated, resulting in a shell of material breaking away from the asteroid, perturbing its orbital velocity. This event is unlikely to fragment the remaining mass of the asteroid. With a surface explosion, a crater is formed and material ejected from the asteroid, perturbing its orbital velocity; however, fragmentation of the asteroid is more likely. Both nuclear approaches for changing the asteroid's orbital velocity require approximately the same explosive energy: 0.01 kiloton, 100 kiloton, and 1 megaton for an asteroid of 100m, 1km, and 10km diameter, respectively. Subsurface nuclear devices could also be exploded to fragment the asteroid; however, these require a landing on the asteroid, drilling to place the device, and a larger explosive. Additionally, it requires more detailed knowledge of the composition of the asteroid and may produce fragments that are still sufficiently large to threaten massive destruction on Earth.[21] We conclude from Ahrens and Harris' work that for the present, a system employing a nuclear device to deflect an Earth-threatening object is feasible.

The other system concepts require significant conceptual feasibility work before one could emerge as a viable long term option. There would understandably be serious concern about the use of a nuclear device in space to fragment or deflect an Earth crossing object due to the potential of radioactive fallout reentering the atmosphere, or damage due to the impact of sufficiently large fragments. While detailed analysis of these risks would need to be performed prior to a decision to employ a nuclear mitigation system against an incoming object, it stands to reason that these liabilities are small in comparison to the danger inherent in failing to mitigate the impact of the object.

Cost

Estimated costs for a NEO surveillance and control network are surprisingly low. The Spaceguard team estimated the cost of their proposed optical network to be $50M investment costs and $8-10M annual operating costs.[22] Bell *et al* estimated the cost of the GEODSS upgrades, including the addition of the Explorer class infrared satellite to be $57M investment cost and $13M annual operating costs. The additional use of the NASA liquid mirror telescopes added approximately $3M per year in operations costs.[23] All these figures include the cost of the control function. Several estimates place the cost of a mitigation system around $1 billion[24]—about half the cost of a B-2 bomber.

Do these costs seem like a lot? Let's compare them with some other numbers, but not costs for space systems. We're not looking to buy a surveillance system because we want to be space pioneers, but because we're trying to avert a natural disaster. The US cost for disaster relief dwarfs the cost of a planetary defense system, even if the above figures are underestimated by an order of magnitude. In June 1997, the Senate passed a disaster relief bill which would provide $5.4 billion for domestic disaster relief for fiscal year 1998 alone.[25] Twenty-five weather-related disasters in the US from 1988 to 1997 had total damages/costs of $140 billion, and 21 disasters between August 1992 and May 1997 cost over $90 billion and 911 deaths.[26] Even "familiar" natural threats are *very* expensive.

Any money provided by the US to fund planetary defense activities must be "new money", as NASA has no current plans to spend its current budget tracking NEOs[27] and planetary defense is not a mission assigned to the Department of Defense or the Air Force,[28] although Air Force Space Command has been tasked to perform a mission area assessment for defense of the planet.[29] New money means Congress provides additional appropriated funds over and above existing agency budgets, or Congress directs an agency to provide the funds "out of hide" to the detriment of some other budgeted project.

Legality of Planetary Defense

Since the NEO and comet threat is spaceborne, defensive measures of necessity must respond to the threat in space. Planetary defense systems therefore must be concerned with public international law, and in particular the subcategory of space law. Space law is primarily based on custom, treaty, and international agreement. Of the three elements of a planetary defense system described above, the one most likely to be impacted by space law is the mitigation system, since it will actively seek to attack the potential impactor in some fashion. Military and civilian surveillance and control systems have long existed in space, and having passive or nondestructive functions they are not likely to arouse legal issues. Three existing treaties which form a major basis for current space law are ripe for exploration with respect to a mitigation system: The Outer Space Treaty, The Nuclear Test Ban Treaty, and The Anti-Ballistic Missile Treaty. Since this paper is not intended to be a legal treatise, the data below summarizes and contrasts the legal analyses of Kunich and Sweet, which are on point for this purpose.[30]

Outer Space Treaty

The United States, the Soviet Union, and more than 100 other nations under United Nations sponsorship signed the Outer Space Treaty in 1967. Generally, the treaty seeks to preserve space for free use and exploration by all nations, to restrict (not eliminate) military activities in space, and to preserve the use of space for peaceful purposes. Through lengthy analysis, Kunich concludes that "a planetary defense system, having as its only target entirely naturalistic forces of nature utterly devoid of human genesis or control, is not a weapon and is not prohibited by the Outer Space Treaty. As with other non-weapons such as a shovel or chisel, some of the components of a planetary defense system, particularly those that could deflect or destroy and asteroid, have a peaceful *purpose*." (emphasis in original)[31] As an analogy, he cites the Strategic Defense Initiative as a system with a peaceful *purpose* (self-defense) even though components of the system could have been used offensively. SDI was viewed as compliant with the Outer Space Treaty; the US continues today with development of ballistic missile defense systems which operate in space.[32] Further, Kunich determines that testing a mitigation system on a celestial body will be permissible since a planetary defense system is not considered a weapon.[33]

Although an advocate of a planetary defense system, which she insightfully labels "planetary preservation" as a more accurate moniker, Sweet disagrees that a mitigation system can not be considered a weapon: "But the argument that the term weapon is applicable solely against entities with some sentient ability is weak and has some flaws. Militaries certainly target adversaries' defensive weapons, populations, and infrastructures."[34] While she is correct that defensive weapons, infrastructures, and other objects are often targets, the inference is not. These are targeted in wartime not as a consequence of their status as inanimate objects, but because of their purpose and use as elements of warfare or sustainment of resistance by their sentient, human masters.

Notwithstanding her stance on the definition of a mitigation system as a weapon, however, Sweet concludes that such a system is permissible under the Outer Space Treaty because the military use of space for "peaceful purposes" has long been accepted. She argues the 1980 Vienna Convention on the Law of Treaties requires the Outer Space Treaty to be interpreted to permit a planetary defense system, since it is not specifically prohibited by treaty. Further, from a "militarization of space" perspective it is consistent with the practices of the spacefaring powers, primarily the United States and the Soviet Union/Russia.[35]

Nuclear Test Ban Treaty

If a mitigation system makes use of a nuclear explosive device as outlined above, then the Nuclear Test Ban Treaty may present a challenge. The treaty's purpose is to prohibit nuclear weapon test explosions, and other nuclear explosions. Since Kunich's argument justifying a planetary defense system as consistent with the Outer Space Treaty was based on a "non-weapon" status, it would seem that the "other nuclear explosion" language would prohibit the use of a nuclear device in a mitigation system. Interestingly, the treaty does not prohibit the use of nuclear weapons in wartime. The US has long maintained, as stated by then-Secretary of State Dean Rusk that the treaty "does not prohibit the use of nuclear weapons in the event of war *nor restrict the exercise of the right of self-defense* recognized in Article 51 of the Charter of the United Nations." (emphasis added)[36]

The Soviet Union gradually adopted a similar interpretation with respect to peaceful nuclear explosions: "It is concluded that, at present, peaceful nuclear explosions are advisable only for *exceptionally urgent problems which cannot otherwise be solved.*" (emphasis added)[37] Certainly use of a mitigation system to deflect or destroy an incoming celestial object would qualify as an exceptionally urgent problem which cannot otherwise be solved, at least until other yet-to-be-developed technologies come to fruition. Kunich concludes that, consistent with the accepted interpretations of the United States and Soviet Union, "neither the testing nor the actual use of a planetary defense nuclear device in space would be precluded by this treaty."[38]

By contrast, Sweet opines that a nuclear mitigation system for the purpose of planetary defense does indeed violate the original intent of the treaty. Based on events which have occurred since ratification of this treaty in the early 1960s; however, such as development of nuclear technology by such countries as China, France, and Iraq, she concludes a nuclear solution to an impending NEO impact is consistent with current interpretation of the treaty.[39]

41

Anti-Ballistic Missile Treaty

The Anti-Ballistic Missile (ABM) Treaty is somewhat more problematic. One of its prohibitions is missiles with the *capability* (as opposed to intended purpose) to counter strategic ballistic missiles in flight. Since we have no system built, this capability can not be ruled out at this time, and the treaty provision could be at issue. Kunich points out, however that the US and Russia have had recent interpretive differences over the treaty, specifically as it relates to ballistic missile defense systems. Further, Congressional sentiment was evident in 1995 and 1996 to amend or withdraw from the treaty for reasons unrelated to planetary defense. As a last resort, Kunich suggests that the President may be forced to unilaterally "withdraw from, terminate, or suspend a treaty" in a true emergency situation such as may be presented in a comet or asteroid impact scenario.[40]

Treaty withdrawal is an alternative in Sweet's analysis, but she generally regards the ABM treaty as an ineffective "means of controlling anything and merely served the political purposes of the superpowers at the time."[41] Further, agreement between the U.S. and Russia in May 1997 that the ABM Treaty permits a theater ballistic missile defense system is undeniable evidence that the ABM Treaty need not be an impediment to a planetary defense system.[42]

The Legal Road Ahead

The extent to which these treaties (all products of the apex of the cold war) retard, interrupt, or diminish efforts to develop an international planetary defense solution, a grave leadership issue exists for the United Nations, and the spacefaring powers in particular. Since international law on this subject is generally based on treaty, any ambiguities, anachronisms, or obfuscational interpretations should be able to be dealt with smoothly by new international agreement, formal or informal. If the legal community deems this necessary for the sake of legal precision or completeness, then rapid action must be taken now. In particular, this action must be *unlike* that currently underway to clarify the use of nuclear power sources in space under the Outer Space Treaty. This activity has been underway since 1978 and to date has resulted only in a recommendation to the United Nations General Assembly.[43] In any case, we must act to ensure that the eventual cost of a planetary defense solution is not eclipsed by the legal costs of enabling it; we certainly don't want to experience an otherwise preventable Earth impact while the legal detailia are debated to the detriment of life on Earth. Every once in a while, common sense must prevail.

Notes

[1] Spaceguard, Executive Summary.

[2] An object need not have an orbit which intersects earth mass to impact the earth. It need only pass close enough to be captured sufficiently by Earth's gravity to alter its orbit to the extent that an impact results; hence, capture cross section.

[3] Spaceguard, sec 1.

[4] Ibid.

[5] Ibid., sec 9.

[6] Ibid., Executive Summary.

[7] Bell, Bender, and Carey, 159.

[8] Ibid., 123-124.

[9] Ibid., 180-184.

[10] Ibid., 196-215.

[11] Spaceguard, sec 5.

[12] Ibid., sec 8.

[13] Bell, Bender, and Carey, 191-192.

[14] Spaceguard, sec 3.

[15] Ibid., sec 6.

[16] Christopher F. Chyba, "Explosions of Small Spacewatch Objects in the Earth's Atmosphere," *Nature* 363, no. 6431 (24 June 1993): 701-703.

[17] Urias et al, sec 3c.

[18] Ibid.

[19] Thomas J. Ahrens and Alan W. Harris, "Deflection and Fragmentation of Near-

Notes

Earth Asteroids," *Nature* 360, no. 6403 (3 December 1992): 430, 433. The velocity increment derived results in deflection of the incoming object by one earth radius. Perihelion is the point in an object's orbit that is nearest the sun.

[20] Ibid., 430-432.

[21] Ibid., 430-433.

[22] Urias et al., sec 9.

[23] Bell, Bender, and Carey, 212. These costs were believed to have a margin of error of plus or minus 25%.

[24] Nici and Kaupa, 102; and Urias, et al, sec 3c.

[25] "Senate Passes Disaster Aid Despite Promised Veto," *Vote Watch*, 5 June 1997, n.p.; on-line, Internet, 22 September 1997, available from http://www.pathfinder.com/ @@g88vvAQAWxMc7B7K/CQ/bills/S19970095.html.

[26] National Climatic Data Center, "Billion Dollar U.S. Weather Disasters 1980-1997," *National Oceanic and Atmospheric Administration*, 17 June 1997, n.p.; on-line, Internet, 29 September 1997, available from http://www.ncdc.noaa.gov/publications/ billionz.html.

[27] Nici and Kaupa, 100.

[28] Spaceguard, Executive Summary; and Bell, Bender, and Carey, 3.

[29] Nici and Kaupa, 100.

[30] The purpose of considering only the analyses of Kunich and Sweet is not to limit or filter the legal perspective on this subject. Indeed, it is significant that these comprise the *only* legal analyses of planetary defense issues.

[31] Kunich, 129-143.

[32] Ibid., 141.

[33] Ibid., 143.

[34] Kathleen Sweet, "Planetary Preservation: A Space Legal Issue Now or a Survival Issue Later," Prepublication article, n.d.

[35] Ibid.

[36] Kunich, 144-148.

[37] Ibid., 148.

[38] Ibid., 149.

[39] Sweet.

[40] Kunich, 150-157.

[41] Sweet.

[42] Ibid.

[43] Ibid.

Chapter 4

Current Programs

One thing is sure. We have to do something. We have to do the best we know how at the moment. If it doesn't turn out right, we can modify it as we go along.

—Franklin Delano Roosevelt

To this point, we've examined the threat to Earth posed by Earth-crossing asteroids and comets by looking at the hazard posed by an impact and the likelihood that an impact will occur. We've posited various methods to evaluate this risk, since the Earth-impact phenomenon involves very low probabilities, yet ultra-high consequences and is outside our collective experience base. We've also seen that the technology is available *today* to address this threat. Let's now look at what's being done to address the NEO threat.

Surveillance Programs

Despite the best efforts of the scientific community and the profound rhetoric of advocacy in the U.S. Congress, virtually no public funding has been made available to augment and accelerate surveillance projects to identify and catalog NEOs. International funding is no more generous. Surveillance efforts are conducted by Spacewatch (University of Arizona), Palomar Observatory, Lowell Observatory, Côte d'Azur Observatory (France), the NASA Jet Propulsion Laboratory in conjunction with the Air

Force, and amateur astronomers. An Anglo-Australian program which had been operational since 1990 was closed by the Australian government on 1 January 1997.[1] Although this seems like a long list of programs, all are small-scale efforts, generally using telescopes of less than 1 meter in diameter. Financially, most are "barely surviving;"[2] as pointed out previously, their progress is agonizingly slow. It is indeed a tribute to the dedication of the scientists involved that they continue to persevere in the face of such scarce funding. So perilous is the funding situation, and so bleak the outlook for public funding that Tom Gehrels, a preeminent expert on the NEO hazard from the University of Arizona, has appealed to the United Nations to endorse the need for increased funding for surveillance programs.[3]

The cooperative effort between the Jet Propulsion Laboratory and the Air Force is called the Near Earth Asteroid Tracker (NEAT). NEAT involves a NASA camera installed on a 1m Air Force GEODSS telescope. In 1996, NEAT operated 12 nights each month; in 1997, the Air Force reduced the surveillance time to 6 nights per month due to "operational impacts" to other Air Force needs.[4] This points to the ultra-low, virtually non-existent priority that planetary defense enjoys within the Air Force.

Characterization Programs

Several ongoing and future programs address the need for a better understanding of the physical characteristics of asteroids and comets[5]. They also demonstrate the increasing international nature of space programs as NASA, the European Space Agency (ESA), Japan, and combined agency programs are represented. From a planetary defense perspective, these programs are far more plentiful and better supported than the surveillance or mitigation elements because the characterization programs are pursued primarily for purposes of planetary exploration and space science. Fortunately, increasing our knowledge of the physical characteristics of comets and asteroids through these programs also supports planetary defense needs. Through them, scientists will better be able to develop effective migitation techniques and better understand the interaction of each type of body with Earth's atmosphere.

Ongoing Programs

The Near-Earth Asteroid Rendezvous (NEAR)—a NASA "Discovery"[6] probe—was launched in February 1996, returned images of the asteroid Mathilde in June 97, and will orbit the asteroid 433 Eros in early 1999.[7] Once in orbit, NEAR will use a battery of instruments to conduct a detailed investigation of the asteroid's surface. This will include a complete mineralogical survey which can be compared to the composition of meteorites found on Earth, determination of the asteroid's shape which will provide information about its internal structure, and photographs which will provide data on inner solar system environmental conditions for the past several billion years.[8]

NASA and ESA launched the Cassini spacecraft on 15 October 1997 with the primary mission of investigating Saturn and one of its moons, Titan. During its journey to Saturn, however, mission planners expect Cassini to be able to make observations of several asteroids as it transits the asteroid belt prior to orbiting Saturn in 2004.[9]

Two Japanese spacecraft, Sakigake and Suisei were launched to flyby Halley's Comet, and continue on extended missions to flyby Comet P/Giacobini-Zinner in November 1998.[10]

Future Programs

NASA's Deep Space 1 craft is the first in its New Millennium Program, and will feature an ion propulsion system. To be launched in July 1998, Deep Space 1 will fly to within 5km of asteroid 3352 McAuliffe the following year, studying its composition, surface features, size, and spin state in as well as its interaction with solar wind. In 2000, it will encounter Comet P/West-Kohoutek-Ikemura, additionally studying the composition of the comet tail. Deep Space 1 will also flyby and conduct similar studies of Mars.[11]

Another project in NASA's Discovery program is the Comet Nucleus Tour (CONTOUR), approved in October 1997. The CONTOUR vehicle will take photographic and spectral images of comet nuclei and analyze the dust (tail) flowing from them. NASA plans to launch CONTOUR in July 2002, encountering comets Encke, Schwassmann-Wachmann-3, and d'Arrest over the period November 2003 through August 2008 at a distance of approximately 100km from the nucleus.[12]

A more extensive NASA project with a mission similar to CONTOUR is the Stardust project. The mission of Stardust is to collect cometary coma samples from the comet Wild-2, passing within 50km of the comet's nucleus. Following a February 1999 launch, Stardust will conduct detailed analyses of the properties of cometary matter, and return particle samples to earth where they can be studied at the highest possible level of detail and sensitivity. Among other things, researchers expect to determine the mineralogical, elemental, and chemical composition of comets and the state of water—ice or liquid—in comets. Additionally, Stardust will take detailed photos of the surface of the comet's nucleus.[13]

Two of the more ambitious characterization missions are planned using the Japanese Muses-C and the ESA Rosetta spacecraft. The objective of Muses-C is to collect and return an asteroid sample to Earth. After a planned launch in January 2002, it will accomplish a soft landing on asteroid 4660 Nereus in September 2003. Surface samples will be collected, and a NASA-supplied rover will also be employed. Muses-C will remain on the asteroid for two months, returning a small capsule containing the samples to Earth in January 2006.[14]

Rosetta has a similar mission to Muses-C in that it will make a soft landing; however, Rosetta is a comet investigator. Sponsored by ESA, Rosetta will be launched in January 2003, and perform flybys of asteroids 3840 Mimistrobell and 2530 Shipka before arriving at Comet P/Wirtanen in August 2011. After orbiting the comet for approximately 1 year, Rosetta will release two probes, RoLand and Champollion to land on the surface and relay scientific data to Earth.[15]

Mitigation Programs

There are no ongoing programs to develop or implement a system to deflect or fragment a potential Earth impactor, although launch vehicles and explosive devices that could form the core of such a system currently exist in the United States, Russia, and possibly other countries. Although not a mitigation system per se, the Clementine II program was a "micro" spacecraft to track and intercept an asteroid in space—tasks a mitigation system would have to perform to perfection. Having intercepted an asteroid, Clementine II would release a probe to impact it would then send back scientific data on the characteristics of the asteroid and the impact ejecta. The spacecraft would have been launched in 1999 to intercept the asteroid Toutatis in October 2000.[16]

Clementine II is described in the past tense because on 14 October 1997, President Clinton used his line-item veto authority to cut 13 programs representing 0.06 percent of the $248 billion defense appropriations bill for fiscal year 1998. One of these 13 minor programs was the $30 million Congress had appropriated for Clementine II.[17] This veto came because the technology to be used for asteroid intercept was originally developed under the Strategic Defense Initiative for intercepting missiles. "The fact that Clementine's research might be applied to shooting down not just wayward asteroids but also enemy missiles was the kiss of death."[18] This appears to be another "interpretation" of the ABM treaty based on political logic—it permits a theater missile defense system but may prohibit a technology demonstration which leverages antimissile technology. And then there were none.

The Air Force

One might expect that the Air Force would be well on the way to implementing a system to address the NEO threat, particularly given NASA's preoccupation with the Space Shuttle and the International Space Station. The mission of the Air Force is "to defend the United States through control and exploitation of air and space."[19] The Air Force says it is "transitioning from an *air* force into an *air and space* force on an evolutionary path to a *space and air* force." (emphasis in original)[20] The Air Force also boasts that its "devotion to air and space power will continue to provide the strategic perspective and rapid response the nation will demand as it enters the 21st Century."[21]

The best measure of an agency's commitment is to look at its budget: is its money where its mouth is? What's in the Air Force budget for planetary defense, now or in the future? Zero. Zip. Nothing. Although a budget upgrading GEODSS to improve its NEO surveillance capabilities was proposed in Air Force Space Command's (AFSPC) long range Space Control Mission Area Plan, it was eventually cut due to low priority.[22]

Although some in the AFSPC leadership seem to feel AFSPC should be leading the national planetary defense effort, they have been reluctant to plan or commit resources without clear definition of an Air Force mission. In fact, the AFSPC level of effort on planetary defense is negligible at this point—the most involved action officer worked it less than 10 percent of his time.[23] If you're expecting the light blue cavalry to ride in to preserve the planet against the natural threat from space, you will be very disappointed.

Notes

[1] Tom Gehrels, "A Proposal to the United Nations Regarding the International Discovery Programs of Near-Earth Asteroids," *Annals of the New York Academy of Science* 822, (1997): 603-605.

[2] Ibid., 603-604.

[3] Ibid., 603-605.

[4] "Near-Earth Asteroid Tracking," *NASA Jet Propulsion Laboratory*. N.d., n.p.; on-line, Internet, 23 October 1997, available from http://huey.jpl.nasa.gov/ ~spravdo/neatintr.html; and Lt Col Lindley N. Johnson, Air Force Space Command, interviewed by author, 24 November 1997.

[5] A number of missions to asteroids and comets have already been completed and have provided scientific data on the nature of these objects. These missions were (1) NASA's International Cometary Explorer (ICE) which flew through the plasma tail of Comet Giacobini-Zinner; (2) the Soviet Vega 1 and Vega 2 craft which, after deploying Venus entry probes, flew by Halley's Comet at distances of 10,000 and 3,000km; (3) ESA's Giotto spacecraft, which was designed specifically for cometary investigation and flew within 596km of Halley's Comet and later within 200km of Comet P/Grigg-Skjellerup; and (4) NASA's Galileo probe performed closeup studies of asteroids Gaspra and Ida while enroute to Jupiter. See National Space Science Data Center, NASA Goddard Space Flight Center, *Master Catalog Display Spacecraft,* n.d., n.p. On-line. Internet, 10 December 1997. Available from http://nssdc.gsfc.nasa.gov/cgi-bin/database and http://nssdc.gsfc.nasa.gov/planetary/asteroidpage.html.

[6] Discovery projects are low cost (generally less than $150 million), rapid implementation projects with high scientific payoff. The most publicly visible program in this series has been the highly successful Sojourner Mars rover. The Discovery program was initiated by NASA Administrator Dan Goldin.

[7] "Missions to Gather Solar Wind Samples and Tour Three Comets Selected as Next Discovery Program Flights," *National Aeronautics and Space Administration,* NASA Headquarters Press Release 97-240, 20 October 1997, n.p.; on-line, Internet, 21 October 1997, available from ftp://ftp.hq.nasa.gov/pub/pao/pressrel/ 1997/97-240.txt.

[8] Jim Bell, "Far Journey to a NEAR Asteroid," *Astronomy* 24, no. 3 (March 1996): 47.

[9] National Space Science Data Center.

Notes

[10] Ibid.

[11] Ibid.

[12] "Missions to Gather Solar Wind Samples"

[13] Donald E. Brownlee et al., "Stardust: Comet Coma Sample Return Plus Interstellar Dust, Science and Technical Approach." *Stardust Home Page*, 21 October 1994, n.p.; on-line, Internet, 21 October 1997. Available from http://stardust.jpl.nasa. gov/sd-mission.html.

[14] National Space Science Data Center.

[15] Ibid.

[16] Briefing, DoD-NASA Clementine Team, subject: Clementine Executive Overview, February 1997.

[17] Eric Pianin and Bradley Graham, "Clinton Tempers Line-Item Approach," *Washington Post*, 15 October 1997.

[18] Editorial, *Wall Street Journal*, 17 November 1997.

[19] Department of the Air Force, *Global Engagement: A Vision for the 21st Century Air Force* (Washington, D.C.: Secretary of the Air Force, n.d.), i.

[20] Ibid., 7.

[21] Ibid., 26.

[22] Johnson. A larger predecessor upgrade to GEODSS to support AFSPC mission requirements other than planetary defense was also not supported.

[23] Ibid.

Chapter 5

Today's Issues

Any danger spot is tenable if men—brave men—will make it so.

—John F. Kennedy

To this point we've examined the threat posed by NEOs and long period comets, including the probability of occurrence and the resultant hazard when an impact occurs; the potential solutions that could be pursued in the near term; and the few, underfunded projects that are now underway. Let's review what we know.

What We Know

- The threat to Earth due to the impact of an Earth-crossing asteroid or comet is not new. The cratering records of the Moon, Mars, and Mercury show impacts have occurred for billions of years. We can find craters on Earth going back hundreds of millions of years, although many have long disappeared due to erosion. Recent history discloses several frightening near misses, and "hits" by bodies too small to cause damage.
- Potential Earth impactors are of two types: Earth-crossing asteroids and short period comets whose orbits can be accurately calculated once the object is discovered, and long period comets, which are generally not detected until they are relatively near Earth. Warning times can range from many years for ECAs to less than a year for long period comets, to zero for any object not detected.
- The hazard created when an object strikes Earth is primarily related to its size, and the variation in the hazard between small and large impactors is extreme. At one extreme, a small impactor could burn up in the atmosphere or explode in the atmosphere with insufficient energy to cause damage. At the other extreme, tens or hundreds of millions of deaths, massive starvation, extinction of species, and an end to civilization as we know it could occur. Any impactor of the proper size at the proper location could conceivably cause any damage in between these two extremes.

- With probability 1.0, Earth will be impacted by comets and asteroids in the future.
- The issue associated with future impact is *when*. No *known* body is on a collision course with Earth, but only 1 percent of potential impactors have been found, and scant progress is being made finding the other 99 percent. An impact could occur tomorrow, or it could occur many years in the future.
- No good barometer of public opinion regarding the risk of a NEO impact exists. Statistically, the risk of dying in a civilization-ending impact is greater than the risk of dying from other "unlikely" events, and is 100 times more likely than sociologists' estimate of the public's threshold of concern.
- Unlike other hazardous conditions with which the human race must contend, the asteroid and comet threat is entirely outside our experience base. This complete lack of experience, coupled with the small statistical probabilities results in a "giggle factor" that colors our ability to objectively analyze the subject.
- The technology exists *today* to identify and track potential threats as well as to deflect those which threaten Earth to a non-threatening trajectory.
- Existing surveillance programs are operating on "shoestring" budgets; but for the perseverance of a handful of dedicated scientists then would probably have already ceased. No program to develop a mitigation system exists. The Air Force is not actively pursuing either. Congress has paid lip service to the threat, but has provided no funding.[1]

Having this information leads not directly to answers, but to questions. Certainly there are countless technical questions, but lacking a program to develop a solution, the relevant questions for today concern policy, responsibility, and priority.

At Issue

Is the asteroid/comet threat of sufficient concern to merit taking action?

Ultimately, one could assert that this is a question of values. My mother taught me that an ounce of prevention is worth a pound of cure. After an asteroid or comet impacts Earth, no cure is available. To wait until the next impact occurs to decide to take action to rectify the problem is far too late—we may not be around to take any action or have any resources with which to act. We claim to live in a culture which values human life, and the United States in particular regularly goes to extraordinary lengths with political, economic, and military action in pursuit of humanitarian objectives. "The development of technology in the past few centuries has been toward increasing understanding and control of natural forces in an effort to improve human life. Protecting ourselves against impacts is a natural extension of those trends, comparable to efforts to develop new drugs and treatments for disease."[2] Do we value human life so little that we reject out of hand—or perhaps with only a giggle—consideration of addressing a problem that threatens not only life and property, but culture and civilization as well? No, our values demand that we take action.

The President's national security strategy defines US national interests in 3 categories: vital interests, important national interests, and humanitarian interests. Vital interests are those of "broad, overriding importance to the *survival, safety and vitality* of our nation." (emphasis added) "We will do whatever it takes to defend these interests. . . ." Important national interests are those which ". . . affect our national well-being and the character of the world in which we live." The United States also acts in the name of humanitarianism. "In the event of natural or manmade disasters or gross violations of human rights, our nation *may act because our values demand it.*"[3] The potential for massive death and destruction resulting from a NEO impact, up to and including the destruction of life and civilization as we know it, fits the definition of all 3 of the President's categories of national interests. It is in our national interest to take action.

But what of those naggingly low statistical figures? According to Willoughby, we are emerging from an "age of ignorance" being blissfully unaware of the NEO threat and unable to defend against it either. We hope to enter the "age of controlled destiny" in which we will have both the knowledge and capability to defend ourselves. In the interim, we are in an "age of blindness and inexcusable incompetence."[4] The accusation of blindness is apt. Having detected only 1 percent of the estimated NEO population, we currently rely on statistical *estimates* of impact frequency to chart our future course rather than relying on *facts* that are within our capability to determine.

Consider also that more than 50 US government programs use threshold criteria to control expenditures based on an investment per life saved. The Environmental Protection Agency and the Occupational Safety and Health Administration use an expenditure threshold of $3 million per life saved; the Federal Aviation Administration's (FAA) policy for evaluating safety alternatives is the expected value of a human casualty from a commercial aircraft crash is $2.6 million.[5] The latter figure is especially important because statistically, the risk of dying as a result of a NEO impact is close to the risk of dying as a result of commercial airline travel.[6] With this obviously well-accepted threshold, less than 400 lives need be saved to justify investing a billion dollars to defend and preserve our planet.

Friedman presents a first order risk management model developed at the University of Southern California to analyze the decision to pursue a planetary defense system of varying capabilities. This analysis examines the NEO threat for the next 200 years, using established probabilities of occurrence and damage expectations. He concludes that based on a methodology minimizing total expected cost (investment cost plus the expected value of damage), the optimum strategy is to intensify the detection activity with radars and space sensors, gather characterization data through rendezvous and inspection missions, and provide for rapid reaction intercept system construction, but without actually constructing the system until a NEO threat has been detected.[7]

Friedman's recommendation to delay actual implementation of a mitigation system assumes that the actual construction can be accomplished within the warning time available. This time, in fact, may not be available, particularly if the first threatening object is a long period comet. Recall also the hypothetical scenario posed earlier concerning the "signal potential" of a NEO-related event as a harbinger of a disastrous future consequence. Once the "giggle factor" is overcome, the risk justifies investment.

Why hasn't action been taken already?

After just having spent $300 to have a brake job done on a car, who wants to spend another $500 to have their timing belt replaced because it *might* break in the next few years, causing over $1000 in damage to the engine? Of course, no one does, even though they know that it will eventually break, but perhaps not soon and perhaps not until after the car is sold. The gamblers among us will decline the investment, while others will take action either in the interest of assured transportation or paying now rather than later. Let's consider why NASA, the Air Force, and others in the government haven't taken action on the asteroid and comet threat, therefore choosing to gamble by their inaction.

NASA's current priorities are making incremental progress on proving the Space Shuttle as an economical launch system, continuing a manned space program on the Russian space station *Mir*, and making the International Space Station a reality. These programs are NASA's lifeblood, and competition for dollars is increasingly fierce in the current budget environment. When additional funds are available, they go to "glamour" space science projects, primarily exploratory spacecraft—those that hopefully can capture the public's imagination—and future projects supporting manned spaceflight such as the spaceplane. In short, NASA has enough budget difficulties in funding what it sees as its "crown jewels," manned spaceflight, space science, and planetary exploration without advocating funds to look for what the media sometimes sensationally brands as "killer asteroids." There's that old giggle factor at work.

The story is much the same in the Air Force. Budget pressures and military downsizing already hamper the Air Force in being able to do what it views as its current job, without adding another task. Additionally, it's a major paradigm shift for a military service. America's armed services defend against our enemies: the Soviet Union, Saddam Hussein, North Korea—"bad guys" with deadly weaponry and malice toward people of freedom. Ancient celestial masses of stone and ice blithely obeying time-honored laws of physics hardly qualify.

What about the Federal Emergency Management Agency (FEMA)? We routinely see them in action when other natural disasters strike. FEMA's mission is to "Reduce the loss of life and property and protect our institutions from all hazards by leading and supporting the Nation in a comprehensive, *risk-based* emergency management program of mitigation, preparedness, response, and recovery." (emphasis added)[8] The agency apparently has decided that asteroid and comet impact risks, together with the mitigation and preparedness action that FEMA could bring to the table are less significant than the investments they make in preparation and response to floods, earthquakes, hurricanes, windstorms, volcanoes, landslides, technological hazards, and fire. This decision was probably made through omission because a risk due to impacts from space objects is not mentioned in FEMA's 10-year strategic plan.

It would seem Congress should have initiated some action to deal with the asteroid and comet threat, and indeed as previously discussed, they did. The problem is they only scratched the surface. After directing two studies over 5 years ago and producing some admirably eloquent rhetoric for the record, they have allowed federally sponsored action to die. Even after the studies and rhetoric, Congress at large is likely as uninformed as the public. Senator McCain said recently "I am cognizant of the very real risk that Earth *may* someday be threatened by a comet or asteroid, but this is a problem *already receiving ample attention* from the scientific community using other federal and private dollars." (emphasis added)[9] The information presented above demonstrates that the attention this threat is receiving is scant at best and inadequate to make meaningful progress, even in the surveillance arena. As former Speaker of the House Tip O'Neill reminded us, "All politics are local." It should be no surprise, then that Congress has not taken action to address an issue that the public has not recognized as one.

Neither should we be surprised that the President has not as yet become a planetary defense crusader. Who's to inform him? The media has yet to bring the threat into focus. Congress, FEMA, NASA, and the Air Force certainly will not involve the President on issues that they themselves ignore.

Here we see a long list of major actors in our government, each of which has at least a passing interest in the subject. Each has elected to do nothing rather because their plate is already full, and because they can. Why can they? Because no one is pushing them. Who can we normally depend on to give our government a gentle nudge when it becomes lethargic? So what of the media? How has it portrayed the asteroid and comet threat, and how has it treated the response?

The media's response has been curious. As each "near miss" (or "near hit") by a passing object has occurred, generally the media has dutifully reported it. Often it is accompanied by a few sound bites of the same information previously presented earlier in this paper. But the synthesis of a problem rarely rings through. The critical questions are not asked of the scientific community, and most importantly of government officials. A discomforting number of reports are afflicted with one or more of the following problems: the report focuses on an arcane detail of the incident, obscuring the true significance; relevant information is presented, although an unsupported but happy conclusion is drawn; or relevant information is presented, yet left completely up to the reader to analyze.

For example, reports of the close passage of Toutatis in December 1992 focused on the unusual shape of the asteroid and the methods NASA used to photograph it.[10] A recent *Washington Post* article subtitled "Celestial Doomsday Rocks Not Imminent, Experts Say" paints, on its surface, a soothing picture of relief—sufficiently soothing that many may not read the article. Yet the information presented in the article hardly supports that picture. NASA scientists are quoted: "We need to know where the other 90 percent [of the larger asteroids estimated to exist] is" and "All this talk of probability is an expression of our ignorance. Either something will hit Earth in 1997 or it won't. Right now, so little of the sky is being scanned that the probability of getting any warning of such an object is zero." The journalist reports "It [an impact causing global damage and disrupting civilization] could happen centuries from now, thousands of centuries from now or next month."[11] How does this support the conclusion that an impact is "not imminent"? This and an article in *Time* imply that the asteroid threat may exist because the end of the Cold War has left the nuclear weapons community without a job.[12]

Most reports provide raw data to let the reader draw conclusions. A *New York Times* article belatedly reporting the passage of asteroid 1989FC recalls another asteroid's close passage in 1937, presents impact damage expectations, and gives a snapshot of 1982 data concerning asteroid deflection.[13] It allows the reader to wonder how prevalent this phenomenon is, why the asteroid was not discovered until 8 days after it passed Earth, and what if anything is being done about future potential impactors. When a close call between two aircraft occurs, the air traffic control system is often severely criticized and the FAA called upon to defend its actions. Who has the media challenged to defend inaction concerning the asteroid and comet threat?

The purpose of the foregoing discussion was not to attack the media. Its purpose was to point out that the normal "watchdog" function the media performs with respect to government is impaired, because planetary defense and the celestial threat is a new experience for the media as well. As presented earlier, this is a difficult subject in which none of us has a relevant experience base for comparison. No citizen, no soldier, no journalist, no politician, no statesman has this experience. That's why leadership must prevail.

Who is responsible for taking action?

The threat to Earth from the potential impact of comets and asteroids is not manmade. And while the United States may be affected either directly or indirectly when an object strikes Earth, the threat to the planet is not uniquely American. We have no existing framework assigning responsibility for Earth defense. Planetary defense could be considered an overarching international responsibility since the threat is universally applicable to all nations. An international agency such as the United Nations could assume a position of planetary leadership to assure the common defense of all nations.

Although pure in theory, such an approach is unlikely to bear fruit.

While there is no demonstrable need for mass hysteria, there indeed must be a sense of urgency to taking action on planetary defense measures. As a lumbering, bureaucratic organization however, the UN is ill suited to rapid action on a complex subject. It can, however, endorse a program initiated by one nation or a partnership among nations as well as sponsor enabling revisions or clarifications to international law. Further, as will be pointed out below, the UN also can play a critical role in equitably apportioning funding responsibilities to all beneficiaries.

The responsibility to step out in front must fall to the United States. The international community looks to the United States to provide global leadership, and the United States is arguably the only nation that can address the total issue. President Clinton has said "Our responsibility is to build the world of tomorrow by embarking on a period of construction—one based on current realities but enduring American values and interests."[14] Fundamental objectives for our Government as set out in the Constitution support this as well: ". . . provide for the common defence, promote the general Welfare, and secure the Blessings of Liberty to ourselves and our posterity. . . ." It is difficult to argue that action to ensure a planetary defense capability runs counter to any of these objectives.

This is not to say that the United States should "go it alone." On the contrary, we must actively seek international partners to assure a high confidence, robust technical solution at an appropriate cost, because the United States is not the sole proprietor of efficient space technology. For example, both Russia and the European Space Agency have demonstrated spacelift capability that is far more economical than American lift.

Who gets the job within the US Government? The two primary American spacefaring agencies are NASA and the Air Force. Most people think of NASA first when they think of space; however, NASA's mission continues to be primarily one of science and space exploration. The President's National Security Strategy, from which military strategies and requisite capabilities are drawn, has as its objective the protection of our nation's fundamental and enduring needs: protecting the lives and safety of Americans; maintaining the sovereignty of the United States, with its values, institutions, and territory intact; and providing for the prosperity of the nation and its people.[15] It's reasonable to assume that when this was written, the "protection" was required against some threatening nation, menacing terrorist organization or other traditional "bad guy." But despite a decided lack of malicious ideology and adhering only to the laws of physics, the natural threat from asteroids and comets can still threaten the lives and safety of Americans (and others), threaten American values and institutions, and devastate the prosperity of not only the nation, but the globe. Planetary defense is a legitimate mission for the armed forces, and the Air Force in particular, to adopt.

Where will the money come from?

Internationally, all nations stand to benefit from the protection of a planetary defense system. A framework for apportioning global costs among nations equitably already exists in the United Nations based on each nation's share of the world economy and its ability to pay. Shares of the UN regular budget, which pays for its core functions, range from the United States' 25 percent to 96 nations that each pay 0.01 percent.[16] This cost sharing arrangement applied to a planetary defense capability would drastically decrease the financial burden to any one nation. Consider a $2 billion investment spread equally over 10 years. The US share of this expense would be only $50 million per year; the nations with the smallest share would pay only $20 thousand each—certainly a small burden compared to the benefit. Precedent also exists within the UN to share costs based on other criteria. For example, the 5 permanent members of the Security Council[17] are assessed a higher share for peacekeeping operations, with other industrialized countries paying a share slightly higher than their share of the regular budget and the least developed countries paying significantly less.[18]

In the United States, the defense budget alone presents many opportunities for funding a planetary defense investment. In this era of "downsizing" or "rightsizing" our armed forces in response to the end of the Cold War, we have gone through many lengthy studies to determine the their proper alignment—the Base Force, Bottom-Up Review, Quadrennial Defense Review, and most recently the National Defense Panel. Despite these efforts, we still experience significant funding injections from Congress over and above what the President's Budget requires.

In the 1998 Defense Appropriations Bill, Congress inserted "750 projects, worth more than $11 billion, that the administration had not requested and did not want."[19] Included in that $11 billion increase were several "big ticket" items, among them $720 million for an Aegis cruiser, and $503 million for 8 C-130J aircraft. Based on the estimates above, these two items alone would pay for a planetary defense system, including mitigation, without any international cost sharing. An argument could be made that even though the President did not request the Aegis and C-130Js, they would still be useful to the armed forces. This is probably fair, except that the Air Force will incur still more infrastructure costs to support the C-130Js, since they are substantially different than the other C-130s in the Air Force's fleet. Consider another item, however. Congress also inserted $250 thousand in the defense appropriations bill "to assist in a pilot project to encourage a Hawaiian cruise firm . . . to build two luxury cruise ships and grant it the exclusive franchise to operate among the Hawaiian islands."[20] How are luxury cruise ships critical to national defense?

Even some in Congress view the budget additions as wasteful. Senator John McCain said of the 1998 defense bill "The president has clearly . . . missed an opportunity to eliminate billions of dollars in low-priority, unnecessary and wasteful spending from the defense budget."[21] There can be no doubt that even as the US downsizes its armed forces, many opportunities exist to make better use of the funds we have if we can lift ourselves out of the politics as usual mindset. Similar opportunities surely exist elsewhere in the federal budget.

What must be done now?

In a nutshell, it's not a question of *if* an impact will occur, it is solely a question of *when*. We must have no delusions of invulnerability. Most importantly, the potential for massive death and destruction, up to and including the destruction of life and civilization as we know it merits immediate action by the United States, acting in international partnership or alone if necessary to develop the capability to defend our planet and its fragile inhabitants from the natural threat of celestial impact. Further, the technology for the surveillance system needed to identify potential threat objects is well in hand and is, in fact, relatively inexpensive. We have the tools available to assemble and integrate a mitigation system to deflect an incoming object. This solution is of necessity a nuclear option. However, given the high stakes of the game—higher than anything ever considered in the history of the human race—the risks associated with a nuclear device are well warranted. As technology advances and other mitigation options mature, consideration and development of more advanced techniques can certainly be pursued.

The issues before us now are simple. First, existing capability to perform the surveillance required to identify and track potential threatening NEOs and long period comets is woefully inadequate. Second, we are unable to respond to a threat with mitigation action.

The United States has long been looked to for global leadership, both as a protector of peace and a pioneer in the peaceful exploration and exploitation of space. Although on the surface, this is an issue of preserving our planet, our civilization, and ourselves, the immediate issue is one of vision and of leadership for the future. It will take moral courage for US and world leaders to step out and declare "Yes, in the name of posterity, *I* will make a commitment that I could easily delay for a year or for a decade. The preservation of our home begins here." Recall again the epigraph of this chapter. President Kennedy said "Any danger spot is tenable if men—brave men—will make it so." So it is with planetary defense. We can continue to live without fear of the NEO threat not by ignorance but by action *if* brave men, men of vision and moral courage, will make it so.

Notes

[1] To be fair, no executive branch agency has requested funds in the President's Budget.

[2] David Morrison and Edward Teller, "Future Issues," in *Hazards Due to Comets and Asteroids*, ed. Tom Gehrels (Tucson, Ariz.: The University of Arizona Press, 1994), 1137.

[3] The White House, *A National Security Strategy for a New Century*, May 1997, 9.

[4] Friedman, 729.

[5] Ibid., 727.

Notes

[6] At first glance, it may seem impossible that death risk by commercial air disaster and by asteroid impact are approximately the same, since hundreds of people die annually in airline crashes, yet death by asteroid is undocumented for hundreds of years. Statistically, the difference is the number of deaths from an airline crash is relatively low, in the low hundreds, while the expected number of deaths from a large asteroid impact is many orders or magnitude higher. Over a long period of time, say a lifetime, a less likely event producing an extremely large number of deaths can result in the same risk as a more likely event producing a small number of deaths.

[7] Ibid., 727-732. One questionable aspect of the Friedman's modeling is the implementation cost of a mitigation system. He assumes that the cost will vary as the cube of the threatening object's diameter since the energy required to deflect the object varies in this way—in other words, more energy requires proportionally more investment. This appears to ignore the existence of nuclear devices of sufficient energy today, and the existence of sufficient lift capability to deliver the device to the threatening object. Certainly this makes Friedman's investment costs very conservative and potentially justifies additional action at the same investment.

[8] Federal Emergency Management Agency, *Strategic Plan FY 1998 through FY 2007: Partnership for a Safer Future,* 30 September 1997, 5, on-line, Internet, 14 January 1998, available from http://166.112.200.140/library/spln_1.htm.

[9] Senator McCain. "Department of Defense Appropriations Act, 1998—Conference Report. (143 Cong. Rec. S9948; Date: 9/25/97). Text from: Congressional Record. Available from: Congressional Compass (Online Service). Bethesda, MD: Congressional Information Service.

[10] William J. Broad, "NASA Photographs an Asteroid Giving Earth a Close Shave, Sort Of," *New York Times,* 4 January 1993. "NASA Gets 1st Clear Image of an Asteroid That Could Someday Imperil Earth," *Washington Post*, 4 January 1993.

[11] Kathy Sawyer, "The Sky Is Falling But Most Pieces Miss: Celestial Doomsday Rocks Not Imminent, Experts Say," *Washington Post,* 16 February 1997.

[12] "Talk About Star Wars," *Time* 139, no. 14 (6 April 1992): 25.

[13] Warren E. Leary, "Big Asteroid Passes Near Earth Unseen In a Rare Close Call," *New York Times,* 20 April 1989.

[14] The White House, *National Security Strategy*, i.

[15] Ibid., 5.

[16] "United Nations General Information," *United Nations Home Page*, n.d., n.p.; on-line, Internet, 1 December 1997, available from http://www.un.org.geninfo/ir/ch3/ch3_txt.htm.

[17] China, France, Russia, the United Kingdom, and the United States

[18] Ibid.

[19] John M. Broder, "Clinton Gently Vetoes Items in Military Budget," *New York Times*, 15 October 1997.

[20] Pianin and Graham.

[21] Ibid. The quote refers to the missed opportunity to use the President's line-item veto.

Chapter 6

Recommendations

If some day an asteroid does strike the Earth, killing not only the human race but millions of other species as well, and we could have prevented it but did not because of indecision, unbalanced priorities, imprecise risk definition and incomplete planning, then it will be the greatest abdication in all of human history not to use our gift of rational intellect and conscience to shepherd our own survival, and that of all life on Earth.

—AIAA Board of Directors

With the foregoing discussion of the threat posed by Near-Earth Objects and long period comets, and with understanding of potential solutions available to be acted upon with current technology, I offer seven recommendations to guide the actions of the United States in responding to the natural threat to Planet Earth.

1 The United States must make an unequivocal top down policy statement supporting implementation of a planetary defense system with international collaboration.
2 Congress and the Administration must make a solid, long term commitment to program objectives and funding to assure program emphasis and stability.
3 The United States must mobilize the international community to support a planetary defense system.
4 The planetary defense effort must remain focused on a single objective—to protect Earth from asteroid and cometary impacts.
5 The planetary defense effort must be structured to properly emphasize its major elements: surveillance and control, mitigation, and new technology.
6 The Air Force must define Near-Earth Object mitigation as an operational national defense space mission.
7 The Air Force must begin work leading to a test of a prototype mitigation system.

National Policy

Recommendation 1: *The United States must make an unequivocal top down policy statement supporting implementation of a planetary defense system with international collaboration.* In the post-Apollo era, the US has been reluctant or unable to set out a clear statement of space policy supported by funded programs that are sharply focused to achieve policy objectives. In the past, US space policy has been driven by foreign policy, not by specific goals it wanted to achieve in space.[1] This has resulted in fragmented efforts at home and confusion among our international partners, collaborators, and customers. Since the Earth impact threat is real, omnipresent, and equally threatens every living being on Earth, international partners in the effort are absolutely necessary. Our policy should be to rapidly implement a robust planetary defense system, including a near term response capability. Further, our policy must be to *actively and sincerely* seek international partnership from the spacefaring nations and support from the international community.

A note of caution: We must assure that an attempt to state a policy directed at pursuing a planetary defense system does not merely result in a "study" to determine that policy. Since 1986, there have been at least 1 high level space policy study conducted each year to address the space launch policy of the US; as a result of all these studies, our policy is substantially unchanged since 1985.[2] If we fall back into the comfort of studies, 10 years from now, and then 20 years from now we'll still be asking the same questions, assuming we're still here to ask them.

At the risk of redundancy, the planetary defense question is one of vision and of leadership. Our leaders must make the decision now—there is more than ample evidence to do so—and task the international military-academic-governmental-industrial space complex to move out. If you don't know where you want to go, no study can chart your course. We know where we must go, and know the road to get there. We need only begin the journey.

Commitment

Recommendation 2: *Congress and the Administration must make a solid, long term commitment to program objectives and funding to assure program emphasis and stability.* We cannot afford for a planetary defense system to become a political football as so many of our other programs have—the stakes are too high. Once we clear the initial hurdle demonstrating the vision to begin, we must stay the course and see it through to an operational system.

Since the government is now experimenting with various "reinvention" programs in pursuit of improved business practices such as the vice president's National Performance Review and many "acquisition reform" initiatives, let us take a bold step for a program that demands audacity. The Planetary Defense program should be authorized and appropriated for a 10 year period—call it a pilot program if you wish—with rigidly fenced funding and clear, concrete program objectives. This will give the leadership of the program clear boundaries within which to execute the program and will assure our partners and supporters of our unwavering commitment. I also assert that it will result in a success story that we will wish to emulate in the future.

This type of arrangement is not without precedent. The recently enacted Amtrak Reform and Accountability Act of 1997 authorizes approximately $1 billion per year to be appropriated to Amtrak for capital equipment and operating expenditures over a 5 year period, for each fiscal year 1998 through 2002.[3]

Further, a strong commitment is required because we must show international partners that our moral and financial commitment is for a much longer period than just our annual appropriation cycle. The turbulence associated with annual appropriation has caused great frustration with cooperative US-ESA space programs in the past. Once the European Space Agency approves a program, the participating governments fund them for the life of the program, although significant cost growth may subject the program to additional review and approval to continue in the face of the cost growth.[4]

For example, in 1981 the US canceled its spacecraft which was part of the NASA-ESA International Solar Polar Mission[5] due to reductions in the NASA budget. "The incredulity regarding NASA's willingness to cancel an international program reflected ESA's stunned realization of a fundamental difference in attitude between the two organizations concerning the sanctity of international agreements."[6] A solid, long term commitment by the US will assure the international community that the planetary defense system will not be a repeat of the Solar Polar embarrassment and promote international technical and financial investment.

International Mobilization

Recommendation 3: *The United States must mobilize the international community to support a planetary defense system.* As discussed previously, the United States is not uniquely threatened by NEOs. Although we could act alone to shoulder the entire burden, it is neither in our interest, nor the interest of the international community for the United States to do so.

We must rapidly mobilize the international community to act in four areas. First, we

71

must assure that international law is clear and unambiguous (if it is possible for law to be so) in permitting construction, testing, and operation of a NEO mitigation system for the purpose of protecting Earth. Although an argument exists that treaty revisions are not required, clarity is always best; it will be some time before anyone is ready to test or operate a system. Second, we need technical cooperation in all three elements of the system: surveillance and control, mitigation, and new technology. Third, we need the United Nations to develop advocacy and a cost allocation basis to equitably share the cost of this global benefit. In the far term, this should probably result in the establishment of an international oversight panel reporting to the UN.

The fourth international action certainly will be the most difficult, but also the most important. We must initiate international discussions to focus on the decision process governing when to employ a mitigation system. Who will make the ultimate decision, and with what consultation? Any decision to use a system to deflect an incoming object will be very profound, but also will likely be highly time-constrained. This will dictate a clear, but very streamlined process. Agreement on the process will take a very long time, possibly even longer than the time to develop and implement the mitigation system. It must not be delayed.

Focus

Recommendation 4: *The planetary defense effort must remain focused on a single objective—to protect Earth from asteroid and cometary impacts.* This effort has one and only one purpose: to protect the inhabitants of Earth from the dangers of natural phenomena—period. In spite of current cost effectiveness emphasis on dual-use technologies, great care must be taken to have no other purpose for elements of the system other than planetary defense. For example, if an orbiting sensor can be used both for NEO detection and tracking and for tracking other objects such as satellites, ballistic missiles, or reentry vehicles for military purposes (offensive or defensive), then the system must be prevented from accomplishing those military functions. Even if this means building and employing a duplicate system at double the cost for the legal military purpose, it *must* be done. Part of the reason for the cancellation of the Clementine II asteroid intercept program was its dual purpose of supporting not only planetary defense technology (asteroid intercept), but also ballistic missile defense technology (missile intercept). The issue here is to assure that a system that may be the single most significant creation since Genesis is not derailed by the suspicions, paranoia, or ill intentions of man to become a modern Tower of Babel.

Proper focus will also assure public awareness and support. "Efforts at fostering greater public awareness [of space programs] flounder on the fact that much of the public is simply uninterested, since the public perceives space as having no direct connection to their lives."[7]

> At the 1994 Goddard Space Symposium in Washington, DC, a member of the audience questioned a panel composed of various space analysts as to their opinions on the following question. If the public were given a choice, would they choose to either maintain support for current NASA activities or forego the current program in favor of a Mars mission? Answers ranged from the expected NASA sales pitch that the public was proud of the American space program and would want it to continue, to a skeptical quip that the public would eagerly trade everything NASA was doing for a Mars mission simply because beyond an occasional shuttle launch the public does not have a clue what NASA is doing and a Mars mission is a simple, easy to visualize and understandable goal.[8]

That the public demands a clear goal before giving its support is obvious. Witness the public support of the Mars Pathfinder/Sojourner project in 1997—a "CNN sound bite" quality goal to land a cheap probe on Mars looking for signs of life, and clear evidence of progress. Planetary defense must exhibit the same focus with clear goals and demonstrable progress.

But we must not only define the focus, we must assure that the program, its advocacy, and its resources *stay* focused. Handberg *et al* have concluded that "sustained presidential commitment is critical for the generation and sustaining of the requisite bureaucratic and congressional support for a successful space program. . . ."[9] Following the Kennedy vision for a manned moon landing, he tasked Vice President Johnson with keeping the program and its support on track. Perhaps that strategy is again appropriate to sustain the focus for planetary defense.

Organization

Recommendation 5: *The planetary defense effort must be structured to properly emphasize its major elements: surveillance and control, mitigation, and new technology.* Surveillance and control responsibility should be vested in the Minor Planet Center; responsibility for the mitigation program should be directed to the Air Force; and a new technology program should be jointly the responsibility of NASA and the Air Force, with NASA as the lead. Each must have a *single individual* empowered to direct the program.

Referring back to the candidate organizations proposed for the surveillance and control task, the Minor Planet Center is recommended to lead this effort because the surveillance aspect of planetary defense is most immediate, requires the greatest degree of international cooperation and the widest spectrum of partners. Vesting the program in a non-governmental agency should facilitate these relationships. Further, the surveillance program is primarily a mission of science. Our short history of space exploration has shown us that we get the most rapid results when give our scientists a few dollars unencumbered by any agenda other than the pure scientific objectives. The space science community has a proud history of international cooperation, great progress with limited funds, and results even in the face of bureaucratic impediments.[10]

The Air Force is recommended to direct implementation of a near term (nuclear) system to mitigate an incoming celestial object since the majority of the technology and nuclear operations experience resides there. Clearly, to use military terminology, this must be a "joint" program—joint with NASA, and joint with Russia. Not only is the United States now partnered with Russia in manned spaceflight via the space station *Mir*, but as the two largest nuclear powers, collaboration will assure the rest of the world that our planetary defense venture will not digress into a nuclear weaponization of space. A private organization, the Space Shield Foundation, has already been organized in Russia to promote scientific research and technology development on hazards due to asteroids and cometary impacts with Earth.[11] We should also find Russia's experience and current capabilities in heavy lift boosters to greatly benefit the program.

NASA should lead the effort to focus technology for eventual replacement of the near term mitigation system and to better characterize the physical characteristics of asteroids and comets. Great potential for commercial and international cooperation exists here. Commercial ventures are already ongoing to study the makeup of asteroids— witness the Near Earth Asteroid Prospector, a private spacecraft targeted to land on an asteroid for scientific investigation, with an eye toward commercial exploitation.[12]

Defense Mission

Recommendation 6: *The Air Force must define Near-Earth Object mitigation as an operational national defense space mission.* America's armed forces are at a crossroads.

For nearly 50 years, and for the entire existence of the Air Force, the focus of the armed services' missions, doctrine, and force structure was to deter and win, if necessary, an armed conflict with the Soviet Union. Alas, that "evil empire" has vanished, and with it went the Cold War and the single, focused threat around which American military might was structured. All services now grapple with downsizing, as well as with non-traditional missions now collectively called operations other than war. NEO mitigation must be a new mission for the Air Force, and a key element of space control.

When the nation pursues a planetary defense program, the Air Force will greatly influence the success of the NEO mitigation element by how it treats the mission internally. Air Force Space Command currently views planetary defense as barely an additional duty, and a poor one at that. The little emphasis it has put on it has been in the area of surveillance, not mitigation. After many years, the "space force" within the Air Force is only now emerging from the shadow of fixed wing, air-breathing operations. The importance of the NEO mitigation mission demands that the Air Force treat the program as if the nation's survival depends on it, because it does. Proper emphasis from the Chief of Staff, and proper recognition and staffing of a new operational mission should fill the bill.

The Air Force must craft its NEO mitigation program with great care, empowering a single individual to implement the system. The Defense Department is often prone to creating positions with lofty titles, but with no clout. A good example is the "space architect," an Air Force general officer position instituted by the Deputy Under Secretary of Defense (Acquisition and Technology) to assure all services' space requirements are met. The position is ineffective because its real responsibility is only that of a coordinator, having no capability to direct action, and holding no purse strings.[13] It must be clear to the Air Force, the Department of Defense, and NASA that the individual assigned to direct implementation of the NEO mitigation system is not a facilitator or coordinator, but is the sole individual charged with directing the program.

System Test

Recommendation 7: *The Air Force must begin work leading to a test of a prototype mitigation system.* Having the technology to mitigate an incoming comet or asteroid does not mean that we know with certainty how to deflect the object successfully. The theoretical underpinning of how to deflect a potential impactor is established; and the propulsion, guidance, control, and explosive technology exists today. However, no one has ever intercepted an object outside Earth orbit with the accuracy that will be required to properly position the nuclear explosive. The cancelled Clementine II spacecraft would have demonstrated this capability. More importantly, no one has ever detonated an explosive device on an asteroid to demonstrate the theoretical result will be achieved. In short, while we know it can be done, we haven't done it. There is still much data to be collected before we are confident that we can deflect an incoming object with the necessary surety.

Dr Edward Teller urges that "experimentation should not be delayed except for strong reasons, since procedures for protection need to be decided on the basis of data on comets and asteroids, part of which can be obtained only through experimentation."[14] He also maintains that "we can give an absolute guarantee that we will have no detectable radioactivity on the Earth" as a result of such tests by intercepting the asteroid as its orbit carrys it *away* from Earth.[15]

A planetary defense program that delays mitigation testing until a specific threatening object is located is a dangerous gamble. It assumes that the object will be identified many years prior to impact so that the mitigation system can be developed, tested, and refined prior to its use on the threatening object. This assumption is fatally flawed for two reasons. First, the optimum time to deflect the object is when it is farthest from Earth—less energy is required to deflect it, and more than one attempt may have to be made should the first attempt not be completely successful for any reason. Any time spent in development after an accurate trajectory for the potential impactor is established dramatically increases risk. The object gets closer each second. The mitigation system's job gets more difficult each second.

Second, the necessary development time simply may not exist. Warning time for an incoming long period comet will never be very long, generally a year at best. Warning time for an Earth-crossing asteroid *can* be many years, but until a robust surveillance system has been implemented and operated for many years, we have little assurance that we won't locate the asteroid until it is very near impact. We must *know* that, when the time comes, we have the solution waiting. We can only know that by collecting the necessary data and testing a system now.

Some maintain that a mitigation system should not be pursued until a specific need arises because it could be used as a weapon. The most notable advocate of this position is probably the late Carl Sagan. "Premature development of any asteroid orbit-modification capability, in the real world and in light of well-established human frailty and fallibility, may introduce a new category of danger that dwarfs that posed by the objects themselves."[16] Certainly this must be considered, and no system incorporating a nuclear explosive should ever be treated cavalierly. Even Sagan, however acknowledged a need for testing, saying "And it might not be too soon to start practicing getting to these worldlets [asteroids] and diverting their orbits, should the hour of need ever arrive."[17]

One could doubt the ultimate benevolence of the United States, whether acting alone or in international partnership. Until 1949, the United States was the world's sole nuclear power. It did not take advantage of this status then for hegemonic gain. Since 1992, the United States has been widely regarded as the world's sole superpower following the collapse of the Soviet Union. It has not taken advantage of this status now for hegemonic gain. If we are to be concerned about a nation's misuse of planetary defense system for national gain, it would seem that it would be far simpler—not to mention quicker and cheaper for that matter—just to use existing weaponry, nuclear or conventional, for that purpose. If we are to be concerned about a rogue individual, we can take solace in the failsafe procedures developed over the course of the last 50 years to prevent such an incident. We should be far more concerned with existing issues of the proliferation of weapons of mass destruction.

If we are to act for self-preservation, we must implement a mitigation system. To see it work for the first time when we need it most is foolhardy. We must all be from Missouri in this regard, insisting that the theory, the concepts, the data, and the hardware be tested as quickly as possible.

Notes

[1] Joan Johnson-Freese and George Moore, "Clash of the Titans of Space Policy," *Nature* 366, no. 6454 (2 December 1993): 402.
[2] Joan Johnson-Freese and Roger Handberg, *Space, The Dormant Frontier: Changing the Paradigm for the 21st Century* (Westport, Connecticut: Praeger Publishers, 1997), 23.
[3] Amtrak Reform and Accountability Act of 1997, Public Law 134, 105th Cong., 1st

Notes

sess. (2 December 1997), sec. 301(a). The funding authorized ranges from $1.138 billion in FY 1998 to $0.955 billion in FY 2002.

[4] Joan Johnson-Freese, "Canceling the US Solar-Polar Spacecraft" *Space Policy* 3, no. 1 (February 1987): 33.

[5] ESA was also providing a spacecraft of different capability to the mission. Both were required to achieve the scientific objectives.

[6] Johnson-Freese and Handberg, 28.

[7] Roger Handberg, Joan Johnson-Freese, and George Moore, "The Myth of Presidential Attention to Space Policy," *Technology in Society* 17, no. 4 (1995): 338.

[8] Johnson-Freese and Handberg, 41.

[9] Handberg, Johnson-Freese, and Moore, 346.

[10] Johnson-Freese and Handberg, 58.

[11] "Space Shield Foundation General Information," *Space Shield Foundation Home Page,* n.d., n.p.; on-line, Internet, 23 January 1998, available from http://www. ch70.chel.su/town/spshf/general.html. The Space Shield Foundation (SSF) organized international conferences dealing with this subject in 1994 and 1996: Conference on Space Protection of the Earth—94 (SPE-94) and SPE-96. Abstracts of the more than 150 papers presented at these conferences are available from the SSF home page.

[12] "Buck Rogers, CEO," *Scientific American* 272, no. 9 (September 1997): 34.

[13] Johnson-Freese and Handberg, 53-55.

[14] Morrison and Teller, 1142.

[15] Gregory H. Canavan, Johndale C. Solem, and John D. G. Rather, *Proceedings of the Near-Earth Object Interception Workshop*, LA—12476-C (Los Alamos, New Mexico: Los Alamos National Laboratory, February 1993), 277.

[16] Carl Sagan and Steven J. Ostro, "Dangers of Asteroid Deflection," *Nature* 368, no. 6471 (7 April 1994): 501.

[17] Canavan, Solem, and Rather, 36.

Chapter 7

Our Legacy for the Future

I believe this nation should commit itself. . . .

—John F. Kennedy

We live in an active universe. Although normally transparent to us, Earth moves through the blackness of space along with billions of other bodies. The history of our home planet shows us that this motion is not always harmonious, as Earth has been struck in the past by objects large and small. Someday—as early as tomorrow but perhaps not for tens or hundreds of years—another heavenly body will collide with Earth. When it happens, this occurrence will not be the last; it will happen again and again. We may not wonder *if* a collision will happen, but only *when* it will happen. If we are very lucky, it will explode in the atmosphere as many have done before and we will heave a sigh of relief. If we are unlucky, we may experience the biggest cataclysm in the history of man, possibly going the way of the dinosaurs before us. It may be that it won't happen to us, but it will happen to our heirs. "And the spectacle haunts us because it seems to carry allegorical import, like the whispery omen of a hovering future."[1]

We have a choice before us. We can act now for our common defense, for the preservation of our planet and our posterity, providing a beneficent legacy for a certain future. Or we can rely on the grace of Newtonian motion. Let us choose the former. And let us choose *now* so that we, our children, and their children will cease being haunted by this ubiquitous peril from the cold blackness of space, forevermore being able to look to the heavens seeing blue skies and starry nights, never having to be afraid.

Notes

[1] "The Talk of the Town," *The New Yorker* 60, no. 53 (18 February 1985): 29. Reference is to the 1994 tragedy at Bhopal, India.

Glossary

ABM	Anti Ballistic Missile
AIAA	American Institute of Aeronautics and Astronautics
CEO	Chief Executive Officer
CNN	Cable News Network
ECA	Earth-Crossing Asteroid
FAA	Federal Aviation Administration
FEMA	Federal Emergency Management Agency
GEODSS	Ground-based Electro-Optical Deep Space Surveillance System
km	Kilometer(s)
m	Meter(s)
NASA	National Aeronautics and Space Administration
NEO	Near-Earth Object
SDI	Strategic Defense Initiative
TNT	Trinitrotoluene (dynamite)
WIRE	Wide-Field-of-View IR Explorer

Bibliography

Ahrens, Thomas J., and Alan W. Harris. "Deflection and Fragmentation of Near-Earth Asteroids." *Nature* 360, no. 6403 (3 December 1992): 429-433.

Air University. Spacecast 2020. Vol. I. Staff Study, June 1994.

American Institute of Aeronautics and Astronautics. "Responding to the Potential Threat of a Near-Earth-Object Impact." Position Paper, American Institute of Aeronautics and Astronautics, September 1995.

Begley, Sharon. "The Science of Doom." *Newsweek* 120, no. 21 (23 November 1992): 56-60.

Bell, Jim. "Far Journey to a NEAR Asteroid." *Astronomy* 24, no. 3 (March 1996):42-47.

Bell, Larry D., William Bender, and Michael Carey. "Planetary Asteroid Defense Study: Assessing the Responding to the Natural Space Debris Threat." Research Paper ACSC/DR/225-95-04. Maxwell AFB, Ala.: Air Command and Staff College, 1995.

Brownlee, Donald E., et al. "Stardust: Comet Coma Sample Return Plus Interstellar Dust, Science and Technical Approach. *Stardust Home Page*, 21 October 1994, n.p. On-line. Internet, 21 October 1997. Available from http://stardust.jpl.nasa.gov/sd-mission.html.

"Buck Rogers, CEO." *Scientific American* 272, no. 9 (September 1997), 34-35.

Canavan, Gregory H., Johndale C. Solem, and John D. G. Rather. *Proceedings of the Near-Earth Object Interception Workshop.* LA—12476-C. Los Alamos, New Mexico: Los Alamos National Laboratory, February 1993.

Chapman, Clark R., and David Morrison. *Cosmic Catastrophes*. New York: Plenum Press, 1989.

Chapman, Clark R., and David Morrison. "Cosmic Impacts, Cosmic Catastrophes Part 2." *Mercury* 19, no. 1 (January/February 1990): 21-30.

Chapman, Clark R., and David Morrison. "Impacts on the Earth by Asteroids and Comets: Assessing the Hazard." *Nature* 367, no. 6458 (6 January 1994): 33-40.

Chyba, Christopher F. "Explosions of Small Spacewatch Objects in the Earth's Atmosphere." *Nature* 363, no. 6431 (24 June 1993): 701-703.

Chyba, Christopher F., Paul J. Thomas, and Kevin J. Zahnle. "The 1908 Tunguska Explosion: Atmospheric Disruption of a Stony Asteroid." *Nature* 361, no. 6407 (7 January 1993): 40-44.

"Closest Approaches to the Earth by Minor Planets." *International Astronomical Union Minor Planet Center.* N.d., n.p. On-line. Internet, 21 October 1997. Available from http://cfa-www.harvard.edu/iau/lists/Closest.html.

Colorado Springs Gazette Telegraph, 30 October 1996. In Kunich, John C. "Planetary Defense: The Legality of Global Survival." *The Air Force Law Review*, 1997, 119-162.

"Congressional Statements on the Impact Hazard." *NASA Ames Space Science Division,*

15 October 1996, n.p. On-line. Internet, 16 September 1997. Available from http:// ccf.arc.hasa.gov/sst/c_statements.html.

Covey, Curt, et al. "Global Climatic Effects of Atmospheric Dust from an Asteroid or Comet Impact on Earth." *Global and Planetary Change* 9 (1994): 263-273.

Cowen, Ron. "The Day the Dinosaurs Died." *Astronomy* 24, no. 4 (April 1996): 34-41.

"Discovery of a Satellite Around a Near-Earth Asteroid." *European Southern Observatory.* ESO Press Release 08-97, 22 July 1997, n.p. On-line. Internet, 21 October 1997. Available from http://www.hq.eso.org/outreach/press-rel/pr-1997/ pr08-97.html.

Federal Emergency Management Agency. "Strategic Plan FY 1998 through FY 2007: Partnership for a Safer Future." 30 September 1997. On-line. Internet, 14 January 1998. Available from http://166.112.200.140/library/spln_1.htm.

Frank, Louis A., with Patrick Huyghe. *The Big Splash.* New York: Birch Lane Press, 1990.

Friedman, George J. "Risk Management Applied to Planetary Defense." *IEEE Transactions on Aerospace and Electronic Systems* 33, no. 2 (April 1997): 721-733.

Gallant, Roy A. "Journey to Tunguska." *Sky and Telescope* 87, no. 6 (June 1994): 38-43.

Gehrels, Tom. "A Proposal to the United Nations Regarding the International Discovery Programs of Near-Earth Asteroids." *Annals of the New York Academy of Sciences* 822 (1997): 603-605.

_____, ed. *Hazards Due to Comets and Asteroids.* Tucson, Ariz: The University of Arizona Press, 1994.

Handberg, Roger, Joan Johnson-Freese, and George Moore. "The Myth of Presidential Attention to Space Policy." *Technology in Society* 17, no. 4 (1995): 337-348.

Huntress, Wesley T., Jr. Testimony before the Subcommittee on Space; Committee on Science, Space and Technology; US House of Representatives, 24 March 1993, n.p. On-line. Internet, 16 September 1997. Available from http://ccf.arc.nasa.gov/sst/ testimony_03.html.

Ivashkin, V.V. and V.V. Smirnov. "An Analysis of Some Methods of Asteroid Hazard Mitigation for the Earth." *Planetary and Space Science* 43, no. 6 (1995): 821-825.

Jaroff, Leon. "A Shot Across Earth's Bow." *Time* 147, no. 23 (3 June 1996): 61-62.

Johnson, Lindley. Air Force Space Command. Briefing Charts presented to International Space University. Preparing for Planetary Defense, 23 July 1997.

Johnson-Freese, Joan. "Cancelling the US Solar-Polar Spacecraft." *Space Policy* 3, no. 1 (February 1987): 24-37.

_____. "Development of a Global EDOS: Political Support and Constraints." *Space Policy* 10, no. 1 (February 1994): 45-55

Johnson-Freese, Joan, and Roger Handberg. *Space, The Dormant Frontier: Changing the Paradigm for the 21st Century.* Westport, Connecticut: Praeger Publishers, 1997.

Johnson-Freese, Joan, and Roger Handberg. "The Tortoise and the Tortoise." *Space Policy* 7, no. 3 (August 1991): 199-206.

Johnson-Freese, Joan, and George Moore. "Clash of the Titans of Space Policy." *Nature* 366, no. 6454 (2 December 1993): 400-402.

Kunich, John C. "Planetary Defense: The Legality of Global Survival." *The Air Force Law Review* 41, 1997, 119-162.

Legislative Update. Headquarters Air Force Space Command Legislative Liaison, 24 October 1997.

Los Angeles Times, 24 June 1993.

Michel, P., P. Farinella, and C Froeschlé. "The Orbital Evolution of the Asteroid Eros and Implications for Collision with Earth." *Nature* 380, no. 6576 (25 April 1996): 689-691.

"Missions to Gather Solar Wind Samples and Tour Three Comets Selected as Next Discovery Program Flights." *National Aeronautics and Space Administration.* NASA Headquarters Press Release 97-240, 20 October 1997, n.p. On-line. Internet, 21 October 1997. Available from ftp://ftp.hq.nasa.gov/pub/pao/pressrel/1997/97-240.txt.

Morrison, David. "Is the Sky Falling?" *Skeptical Inquirer* 21, no. 3 (May/June 1997): 22-28.

_____. "Target: Earth!" *Astronomy* 23, no. 10 (October 1995): 34-41.

_____. Testimony before the Subcommittee on Space; Committee on Science, Space and Technology; US House of Representatives, 24 March 1993, n.p. On-line. Internet, 16 September 1997. Available from http://ccf.arc.nasa.gov/sst/testimony_01 .html.

National Climatic Data Center. "Billion Dollar U.S. Weather Disasters 1980-1997." *National Oceanic and Atmospheric Administration*, 17 June 1997, n.p. On-line. Internet, 29 September 1997. Available from http://www.ncdc.noaa.gov/ publications/billionz.html.

National Space Science Data Center, NASA Goddard Space Flight Center. "Master Catalog Display Spacecraft." n.d., n.p. On-line. Internet, 10 December 1997. Available from http://nssdc.gsfc.nasa.gov/cgi-bin/database.

"Near Earth Asteroid Prospector." *Space Development Corporation*, n.p. On-line. Internet, 26 September 1997. Available from http://www.spacedev.com/ SpaceDev/NEAP.html.

"Near-Earth Asteroid Tracking." *NASA Jet Propulsion Laboratory*. N.d., n.p. On-line. Internet, 23 October 1997. Available from http://huey.jpl.nasa.gov/~spravdo/ neatintr.html."NEAT Begins Operations." *NASA Jet Propulsion Laboratory*. NASA/JPL Press Release, 24 April 1996, n.p. On-line. Internet, 21 October 1997. Available from http://huey.jpl. nasa.gov/~spravdo/neatpio.html.

NEO News. NASA Ames Research Center, 10 February 1998.

New York Times, 20 April 1989.

_____, 23 April 1989.

_____, 7 April 1992.

_____, 4 January 1993.

_____, 15 October 1997.

Nici, Rosario, and Douglas Kaupa. "Planetary Defense: Department of Defense Cost for the Detection, Exploration, and Rendezvous Mission of Near-Earth Objects." *Airpower Journal* XI, no. 2 (Summer 1997): 94-104.

Powell, Corey S. "Asteroid Hunters." *Scientific American* 268, no. 6 (June 1993): 34-40.

Proceedings Report. Fifth International Conference on Space '96. "Engineering, Construction, and Operations in Space V." American Society of Civil Engineers, New York, 1996.

Proceedings Report. The Eighth National Space Symposium, United States Space Foundation, 31 March – 3 April 1992.

Public Law 134. 105th Cong., 1st sess., 2 December 1997.

Rabinowitz, T., et al. "Evidence for a Near-Earth Asteroid Belt." *Nature* 363, no. 6431 (24 June 1993): 704-706.

Rather, John D. G. Testimony before the Subcommittee on Space; Committee on Science, Space and Technology; US House of Representatives, 24 March 1993, n.p. On-line. Internet, 16 September 1997. Available from http://ccf.arc.nasa.gov/sst/testimony_02. html.

Sagan, Carl, and Steven J. Ostro. "Dangers of Asteroid Deflection." *Nature* 368, no. 6471 (7 April 1994): 501.

"Senate Passes Disaster Aid Despite Promised Veto." *Vote Watch*, 5 June 1997, n.p. On-line. Internet, 22 September 1997. Available from http://www.pathfinder.com/@@g88vvAQAWxMc7B7K/CQ/bills/S19970095.html.

Slovic, Paul. "Perception of Risk." *Science* 236, no. 4799 (17 April 1987): 280-285.

Space Science Division, NASA Ames Research Center. "Spaceguard Survey: Report of the NASA International Near-Earth-Object Detection Workshop." Report Submitted to Congress, 25 January 1992, n.p. On-line. Internet, 16 September 1997. Available from http://ccf.arc.nasa.gov/sst/spaceguard.html.

"Space Shield Foundation General Information." *Space Shield Foundation Home Page,* n.d., n.p. On-line. Internet, 23 January 1998, available from http://www.ch70.chel.su/town/spshf/general.html.

"Spacewatch Discoveries." *Spacewatch Home Page*, n.d., n.p. On-line. Internet 23 February 1998. Available from http://xlr8.lpl.arizona.edu.spacewatch/discoveries2.html.

Spencer, John R., and Jacqueline Mitton, ed. *The Great Comet Crash: The Impact of Comet Shoemaker-Levy-9 on Jupiter.* Cambridge, United Kingdom: Cambridge University Press, 1995.

Starr, Chauncey, and Chris Whipple. "Risks of Risk Decisions." *Science* 208, no. 4448 (6 June 1980): 1114-1119.

Sweet, Kathleen. "Planetary Preservation: A Space Legal Issue Now or A Survival Issue Later." Prepublication article, n.d.

"Talk About Star Wars." *Time* 139, no. 14 (6 April 1992): 25.

Teller, Edward. To John Major, Prime Minister. Letter, 4 November 1996. n.p. On-line. Internet, 17 September 1997. Available from http://dspace.dial.pipex.com/town/terrace/fr77/teller.htm.

"The Talk of the Town." *The New Yorker* 60, no. 53 (18 February 1985): 29-33.

The White House. *A National Security Strategy for a New Century*. May 1997.

Wilson, Richard, and E.A.C. Crouch. "Risk Assessment and Comparisons: An Introduction. *Science* 236, no. 4799 (17 April 1987): 267-270.

"United Nations General Information." *United Nations Home Page*, n.d., n.p. On-line. Internet, 1 December 1997, available from http://www.un.org.geninfo/ir/ch3/ch3_txt.htm.

Urias, John M., et al.. "Planetary Defense: Catastrophic Health Insurance for Planet Earth." Research Paper Submitted to Air Force 2025, October 1996, n.p. On-line. Internet, 23 September 1997. Available from http://www.au.af.mil/au/2025volume3/

chap16/v3c16-1.htm.

Wall Street Journal, 17 November 1997.

Washington Post, 23 April 1989.

_____, 27 April 1989.

_____, 29 December 1992.

_____, 4 January 1993.

_____, 21 June 1993.

_____, 12 December 1994.

_____, 18 May 1996.

_____, 28 November 1996.

_____, 16 February 1997.

_____, 15 October 1997.

Yabushita, S. and N. Hatta. "On the Possible Hazard on the Major Cities Caused by Asteroid Impact in the Pacific Ocean." *Earth, Moon and Planets* 65 (1994): 7-13.